Rough and Tumble Engineering

James H. Maggard

Contents

ROUGH AND TUMBLE ENGINEERING

BY

James H. Maggard

PREFACE

In placing this book before the public the author wishes it understood that it is not his intention to produce a scientific work on engineering. Such a book would be valuable only to engineers of large stationary engines. In a nice engine room nice theories and scientific calculations are practical. This book is intended for engineers of farm and traction engines, "rough and tumble engineers," who have everything in their favor today, and tomorrow are in mud holes, who with the same engine do eight horse work one day and sixteen horse work the next day. Reader, the author has had all these experiences and you will have them, but don't get discouraged. You can get through them to your entire satisfaction.

Don't conclude that all you are to do is to read this book. It will not make an engineer of you. But read it carefully, use good judgment and common sense, do as it tells you, and my word for it, in one month, you, for all practical purposes, will be a better engineer than four-fifths of the so-called engineers today, who think what they don't know would not make much of a book. Don't deceive yourself with the idea that what you get out of this will be merely "book learning." What is said in this will be plain, unvarnished, practical facts. It is not the author's intention to use any scientific terms, but plain, everyday field terms. There will be a number of things you will not find in this book, but nothing will be left out that would be of practical value to you. You will not find any geometrical figures made up of circles, curves, angles, letters and figures in a vain effort to make you understand the principle of an eccentric. While it is all very nice to know these things, it is not necessary, and the putting of them in this book would defeat the very object for which it was intended. Be content with being a good, practical, everyday engineer, and all these things will come in time.

INTRODUCTORY

I f you have not read the preface on the preceding pages, turn back and read it. You will see that we have stated there that we will use no scientific terms, but plain every day talk. It is presumed by us that there will be more young men, wishing to become good engineers, read this work than old engineers. We will, therefore, be all the more plain and say as little as possible that will tend to confuse the learner, and what we do say will be said in the same language that we would use if we were in the field, instructing you how to handle your engine. So if the more experienced engineer thinks we might have gone further in some certain points, he will please remember that by so doing we might confuse the less experienced, and thereby cover up the very point we tried to make. And yet it is not to be supposed that we will endeavor to make an engineer out of a man who never saw an engine. It is, therefore, not necessary to tell the learner how an engine is made or what it looks like. We are not trying to teach you how to build an engine, but rather how to handle one after it is built; how to know when it is in proper shape and how to let it alone when it is in shape. We will suppose that you already know as much as an ordinary water boy, and just here we will say that we have seen water haulers that were more capable of handling the engine for which they were hauling water, than the engineer, and the engineer would not have made a good water boy, for the reason that he was lazy, and we want the reader to stick a pin here, and if he has any symptoms of that complaint, don't undertake to run an engine, for a lazy engineer will spoil a good engine, if by no other means than getting it in the habit of loafing.

PART FIRST

In order to get the learner started, it is reasonable to suppose that the engine he is to run is in good running order. It would not be fair to put the green boy onto an old dilapidated, worn-out engine, for he might have to learn too fast, in order to get the engine running in good shape. He might have to learn so fast that he would get the big head, or have no head at all, by the time he got through with it. And I don't know but that a boy without a head is about as good as an engineer with the big head. We will, therefore, suppose that his engine is in good running order. By good running order we mean that it is all there, and in its proper place, and that with from ten to twenty pounds of steam, the engine will start off at a good lively pace. And let us say here, (remember that we are talking of the lone engine, no load considered,) that if you are starting a new engine and it starts off nice and easy with twenty pounds, you can make up your mind that you have an engine that is going to be nice to handle and give you but little, if any, trouble. But if it should require fifty or sixty pounds to start it, you want to keep your eyes open, something is tight; but don't take it to pieces. You might get more pieces than you would know what to do with. Oil the bearings freely and put your engine in motion and run it carefully for a while and see if you don't find something getting warm. If you do, stop and loosen up a very little and start it up again. If it still heats, loosen about the same as before, and you will find that it will soon be all right. But remember to loosen but very little at a time, for a box or journal will heat from being too loose as quickly as from being too tight, and you will make trouble for yourself, for, inexperienced as you are, you don't know whether it is too loose or too tight, and if you have found a warm box, don't let that box take all of your attention, but keep an eye on all other bearings. Remember that we are not threshing yet, we just run the engine out of shed, (and for the sake of the engine and the young engineer, we hope that it did not stand out all winter) and are getting in shape for a good fall's run. In the meantime, to find out if anything heats, you can try your pumps, but to help you along, we will suppose that your pump, or injector, as the case may be, works all right.

Now suppose we go back where we started this new engine that was slow to start with less than fifty pounds, and when it did start, we watched it carefully and found after oiling thoroughly that nothing heated as far as we could see. So we conclude that the trouble must be in the cylinder. Well, what next? Must we take off the cylinder head and look for the trouble? Oh, no, not by any means. The trouble is not serious. The rings are a little tight, which is no serious fault. Keep them well oiled and in a day or two ten pounds will start the empty engine in good shape. If you are starting an engine that has been run, the above instructions are not necessary, but if it is a new one these precautions are not out of the way, and a great deal of the trouble caused in starting a new engine, can be avoided if these precautions are observed.

It is not uncommon for a hot box to be caused from a coal cinder dropping in the box in shipment, and before starting a new engine, clean out the boxes thoroughly, which can be done by taking off the caps, or top box, and wiping the journal clean with an oily rag or waste, and every engineer should supply himself with this very necessary article, especially if he is the kind of an engineer who intends to keep his engine clean.

The engine should be run slowly and carefully for a while, to give a chance to find out if anything is going to heat, before putting on any load.

Now if your engine is all right, you can run the pressure up to the point of blowing off, which is from one hundred to one hundred and ten pounds. Most new pop valves, or safety valves, are set at this pressure. I would advise you to fire to this point, to see that your safety is all right. It is not uncommon for a new pop to stick, and as the steam runs up it is well to try it, by pulling the relief lever. If, on letting it go, it stops the escaping, steam at once, it is all right. If, however, the steam continues to escape, the valve sticks in the chamber. Usually a slight tap with a wrench or a hammer will stop it at once, but never get excited over escaping steam, and perhaps here is as good a place as any to say to you, don't get excited over anything. As long as you have plenty of water, and know you have, there is no danger.

The young engineer will most likely wonder why we have not said something about the danger of explosions. We did not start to write about explosions. That is just what we don't want to have anything to do with. But, you say, is there no danger of a boiler exploding? Yes. But if you wish to explode your boiler you must treat it very differently from the way we advise. We have just stated, that as long as you have plenty of water, and know you have, there is no danger. Well, how are you to know? This is not a difficult thing to know, provided your boiler is fitted with the proper appliances, and all builders of any prominence, at this date, fit their boilers with from two to four try-cocks, and a glass gauge. The boiler is tapped in from two to four places for the try-cocks, the location of the cocks ranging from a line on a level with the crown sheet, or top of fire box, to eight inches above, depending somewhat on the amount of water space above the crown sheet, as this space differs very materially in different makes of the same sized boiler. The boiler is also tapped on or near the level of crown sheet, to receive the lower water glass cock and directly above this, for the top cock. The space between this shows the safe variation of the water. Don't let the water get above the top of the glass, for if you are running your engine at hard work, you may knock out a cylinder head, and don't let it get below the lower gauge, or you may get your head knocked off.

Now the glass gauge is put on for your convenience, as you can determine the location of the water as correctly by this as if you are looking directly into the boiler, provided, the glass gauge is in perfect order. But as there are a number of ways in which it may become disarranged or unreliable, we want to impress on your mind that you, must not depend on it entirely. We will give these causes further on. You are not only provided with the glass gauge, but with the try-cocks. These cocks are located so that the upper and lower cock is on or near the level with the lower and upper end of the glass gauge. With another try-cock about on a level with the center of glass gauge, or in other words, if the water stands about the center of glass it will at the same time show at the cock when tried. Now we will suppose that your glass gauge is in perfect condition and the water shows two inches in the glass. You now try the lower cock, and find plenty of water; you will then try the next upper cock and get steam. Now as the lower cock is located below the water line, shown by the glass, and the second cock above this line, you not only see the water line by the glass, but you have a way of proving it. Should the water be within two inches of the top of glass you again have the line between two cocks and can also prove it. Now you can know for a certainty, where the water stands in the boiler, and we repeat when you know this, there is nothing to fear from this source, and as a properly constructed boiler never explodes, except from low water or high pressure, and as we have already cautioned you about your safety valve, you have nothing to fear, provided you have made up your mind to follow these instructions, and unless you can do this, let your job to one who can. Well, you say you will do as we have directed, we will then go back to the gauges. Don't depend on your glass gauge alone, for several reasons. One is, if you depend on the glass entirely, the try-cocks become limed up and are useless, solely because they are not used.

Some time ago the writer was standing near a traction engine, when the engineer, (I guess I must call him that) asked me to stay with the engine a few minutes. I consented. After he had been gone a short time I thought I would look after the water. It showed about two inches in the glass, which was all right, but as I have advised you, I proposed to know that it was there and thought I would prove it by trying the cocks. But on attempting to try them I found them limed up solid. Had I been hunting an engineer, that fellow would not have secured the job. Suppose that before I had looked at the glass, it had bursted, which it is liable to do any time. I would have shut the gauge cocks off as soon as possible to stop the escaping steam and water. Then I would have tried the cocks to find where the water was in the boiler. I would have been in a bad boat, not knowing whether I had water or not. Shortly after this the fellow that was helping the engine run (I guess I will put it that way) came back. I asked him what the trouble was with his try cocks. He said, "Oh, I don't bother with them." I asked him what he would do if his glass should break. His reply was, "Oh, that won't break." Now just such an engineer as that spoils many a good engine, and then blames it on the manufacturer. Now this is one good reason why you are not to depend entirely on the glass gauge. Another equally as good reason is, that your glass may fool you, for you see the try-cocks may lime up, so may your glass gauge cocks, but you say you use them. You use them by looking at them. You are not letting the steam or water escape from them every few minutes and thereby cutting the lime away, as is the case with try-cocks. Now you want to know how you are to keep them open. Well, that is easy. Shut off the top gauge and open the drain cock at bottom of gauge cock. This allows the water and steam to flow out of the lower cock. Then after allowing it to escape a few seconds, shut off the lower gauge and open the top one, and allow it to blow about the same time. Then shut the drain cock and open both gauge cocks and you will see the water seek its level, and you can rest assured that it is reliable. This little operation I want you to perform every day you run an engine. It will prevent you from thinking you have water. I don't want you to think so. I intend that you shall know it. You remember we said, if you know you have water, you are safe, and every one around you will be safe.

Now here is something I want you to remember. Never be guilty of going to your engine in the morning and building a fire simply because you see water in the glass. We could give you the names of a score of men who have ruined their engines by doing this very thing. You, as a matter of course, want to know why this can do any harm. It could not, if the water in the boiler was as high as it shows in the glass, but it is not always there, and that is what causes the trouble. Well, if it showed in the glass, why was it not there? You probably have lived long enough in the world to know that there are a great many boys in it, and it seems to be second nature with them to turn everything on an engine that is possible to turn. All glass gauge cocks are fitted with a small hand wheel. The small boy sees this about the first thing and he begins to turn it, and he generally turns as long as it turns easy, and when it stops he will try the other one, and when it stops he has done the mischief, by shutting the water off from the boiler, and all the water that was in the glass remains there. You may have stopped work with an ordinary gauge of water, and as water expands when heated, it also contracts when it becomes cool. Water will also simmer away, if there is any fire left in the fire box, especially if there should be any vent or leak in the boiler, and the water may by morning have dropped to as much as an inch below the crown sheet. You approach the engine and on looking at the glass, see two or three inches of water. Should you start a fire without investigating any further, you will have done the damage, while if you try the gauge cocks first you will discover that some one has tampered with the engine. The boy did the mischief through no malicious motives, but we regret to say that there are people in this world who are mean enough to do this very thing, and not stop at what the boy did unconsciously, but after shutting the water in the gauge for the purpose of deceiving you, they then go to the blow-off cock and let enough water out to insure a dry crown sheet. While I detest a human being guilty of such a dastardly trick, I have no sympathy to waste on an engineer who can be caught in this way. So, if by this time you have made up your mind never to build a fire until you know where the water is, you will never be fooled and will never have to explain an accident by saying, "I thought I had plenty of water." You may be fooled in another way. You are aware that when a boiler is fired up or in other words has a steam pressure on, the air is excluded, so when the boiler cools down, the steam condenses and becomes water again, hence the space which was occupied by steam now when cold becomes a vacuum.

Now should your boiler be in perfect shape, we mean perfectly tight, your throttle equally as tight, your pump or injector in perfect condition and you were to' leave your engine with the hose in the tank, and the supply globe to your pump open, you will find on returning to your engine in the morning that the boiler will be nearly if not quite full of water. I have heard engineers say that someone had been tampering with their engines and storm around about it, while the facts were that the supply being open the water simply flowed in from atmospheric pressure, in order to fill the space made vacant by the condensed steam. You will find further on that all check valves are arranged to prevent any flowing out from the boiler, but nothing to prevent water flowing in. Such an occurrence will do no harm but the knowing how it was done may prevent your giving yourself away. A good authority on steam boilers, says: "All explosions come either from poor material, poor workmanship, too high pressure, or a too low gauge of water." Now to protect yourself from the first two causes, buy your engine from some factory having a reputation for doing good work and for using good material. The last two causes depend very much on yourself, if you are running your own engine. If not, then see that you have an engineer who knows when his safety valve is in good shape and who knows when he has plenty of water, or knows enough to pull his fire, when for some reason, the water should become low. If poor material and poor workmanship were unknown and carelessness in engineers were unknown, such a thing as a boiler explosion would also be unknown.

You no doubt have made up your mind by this time that I have no use for a careless engineer, and let me add right here, that if you are inclined to be careless, forgetful,(they both mean about the same thing,) you are a mighty poor risk for an insurance company, but on the other hand if you are careful and attentive to business, you are as safe a risk as any one, and your success and the durability and life of your engine depends entirely upon you, and it is not worth your while to try to shift the responsibility of an accident to your engine upon some one else.

If you should go away from your engine and leave it with the water boy, or anyone who might be handy, or leave it alone, as is often done, and something goes wrong with the engine, you are at fault. You had no business to leave it, but you say you had to go to the separator and help fix something there. At the separator is not your place. It is not our intention to tell you how to run both ends of an outfit. We could not tell you if we wanted to. If the men at the separator can't handle it, get some one or get your boss to get some one who can. Your place is at the engine. If your engine is running nicely, there is all the more reason why you should stay by it, as that is the way to keep it running nicely. I have seen twenty dollars damage done to the separator and two days time lost all because the engineer was as near the separator as he was to the engine when a root went into the cylinder. Stay with your engine, and if anything goes wrong at the separator, you are ready to stop and stop quickly, and if you are signalled to start you are ready to start at once You are therefore making time for your employer or for yourself and to make time while running a threshing outfit, means to make money. There are engineers running engines today who waste time enough every day to pay their wages.

There is one thing that may be a little difficult to learn, and that is to let your engine alone when it is all right. I once gave a young fellow a recommendation to a farmer who wanted an engineer, and afterward noticed that when I happened around he immediately picked up a wrench and commenced to loosen up first one thing and then another. If that engineer ever loses that recommendation he will be out of a job, if his getting one depends on my giving him another. I wish to say to the learner that that is not the way to run an engine. Whenever I happen to go around an engine, (and I never lose an opportunity) and see an engineer watching his engine, (now don't understand me to mean standing and gazing at it,) I conclude that he knows his business. What I mean by watching an engine is, every few minutes let your eye wander over the engine and you will be surprised to see how quickly you will detect anything out of place. So when I see an engineer watching his engine closely while running, I am most certain to see another commendable feature in a good engineer, and that is, when he stops his engine he will pick up a greasy rag and go over his engine carefully, wiping every working part, watching or looking carefully at every point that he touches. If a nut is working loose he finds it, if a bearing is hot he finds it. If any part of his engine has been cutting, he finds it. He picked up, a greasy rag instead of a wrench, for the engineer that understands his business and attends to it never picks up a wrench unless he has something to do with it. The good engineer took a greasy rag and while he was using it to clean his engine, he was at the same time carefully examining every part. His main object was to see that everything was all right. If he had found a nut loose or any part out of place, then he would have taken his wrench, for he had use for it.

Now what a contrast there is between this engineer and a poor one, and unfortunately there are hundreds of poor engineers running portable and traction engines. You will find a poor engineer very willing to talk. This is bad habit number one. He cannot talk and have his mind on his work. Beginners must not forget this. When I tell you how to fire an engine you will understand how important it is, The poor engineer is very apt to ask an outsider to stay at his engine while he goes to the separator to talk. This is bad habit number two. Even if the outsider is a good engineer, he does not know whether the pump is throwing more water than is being used or whether it is throwing less. He can only ascertain this by watching the column of water in the glass, and he hardly knows whether to throw in fuel or not. He don't want the steam to go down and he don't know at what pressure the pop valve will blow off. There may be a box or journal that has been giving the engineer trouble and the outsider knows nothing about it. There are a dozen other good reasons why bad habit number two is very bad.

If you will watch the poor engineer when he stops his engine, he will, if he does anything, pick up a wrench, go around to the wrist pin, strike the key a little crack, draw a nut or peck away at something else, and can't see anything for grease and dirt. When he starts up, ten to one the wrist pin heats and he stops and loosens it up and then it knocks. Now if he had picked up a rag instead of a wrench, he would not have hit that key but he would have run his hand over it and if he had found it all right, he would have let it alone, and would have gone over the balance of the engine and when he started up again his engine would have looked better for the wiping it got and would have run just as well as before he stopped it. Now you will understand why a good engineer wears out more rags than wrenches, while a poor one wears out more wrenches than rags. Never bother an engine until it bothers you. If you do, you will make lots of grief for yourself.

I have mentioned the bad habits of a poor engineer so that you may avoid them. If you carefully avoid all the bad habits connected with the running of an engine, you will be certain to fall into good habits and will become a good engineer.

TINKERING ENGINEERS

After carelessness, meddling with an engine comes next in the list of bad habits. The tinkering engineer never knows whether his engine is in good shape or not, and the chances are that if he should get it in good shape he would not know enough to let it alone. If anything does actually get wrong with your engine, do not be afraid to take hold of it, for something must be done, and you are the one to do it, but before you do anything be certain that you know what is wrong. For instance, should the valve become disarranged on the valve stem or in any other way, do not try to remedy the trouble by changing the eccentric, or if the eccentric slips do not go to the valve to mend the trouble. I am well aware that among young engineers the impression prevails that a valve is a wonderful piece of mechanism liable to kick out of place and play smash generally. Now let me tell you right here that a valve (I mean the ordinary slide valve, such as is used on traction and portable engines), is one of the simplest parts of an engine, and you are not to lose any sleep about it, so be patient until I am ready to introduce you to this part of your work. You have a perfect right to know what is wrong with the engine. The trouble may not be so serious and it is evident to you that the engine is not running just as nicely as it should. Now, if your engine runs irregularly, that is if it runs up to a higher speed than you want, and then runs down, you are likely to say at once, "Oh I know what the trouble is, it is the governor." Well, suppose it is, what are you going to do about it, are you going to shut down at once and go to tinkering with it? No, don't do that, stay close to the throttle valve and watch the governor closely. Keep your eye on the governor stem, and when the engine starts off on one of its high speed tilts, you will see the stem go down through the stuffing box and then stop and stick in one place until the engine slows down below its regular speed, and it then lets loose and goes up quickly and your engine lopes off again. You have now located the trouble. It is in the stuffing box around the little brass rod or governor stem. The packing has become dry and by loosening it up and applying oil you may remedy the trouble until such time as you can repack it with fresh packing. Candle wick is as good for this purpose as anything you can use.

But if the governor does not act as I have described and the stem seems to be perfectly free and easy in the box, and the governor still acts queerly, starting off and running fast for a few seconds, and then suddenly concluding to take it easy and away goes the engine again, see if the governor belt is all right, and if it is, it would be well for you to stop and see if a wheel is not loose. It might be either the little belt wheel or one of the little cog wheels. If you find these are all right, examine the spool on the crank shaft from which the governor is run and you will probably find it loose. If the engine has been run for any length of time, you will always find the trouble in one of these places, but if it is a new one the governor valve might fit a little tight in the valve chamber and you may have to take it out and use a little emery paper to take off the rough projections on the valve. Never use a file on this valve if you can get emery paper, and I would advise you to always have some of it with you. It will often come handy. Now if the engine should start off at a lively gait and continue to run still faster, you must stop at once. The trouble this time is surely in the governor. If the belt is all right, examine the jam nuts on the top of the governor valve stem. You will probably find that these nuts have worked loose and the rod is working up, which will increase the speed of the engine. If these are all right, you will find that either a pulley or a little cog wheel is loose. A quick eye will locate the trouble before you have time to stop. If the belt is loose, the governor will lag while the engine will run away. If the wheel is loose, the governor will most likely stop and the engine will go on a tear. If the jam nut has worked loose, the governor will run as usual, except that it will increase its speed as the speed of the engine is increased. Now any of these little things may happen and are likely to. None of them are serious, provided you take my advice, and remain near the engine. Now if you are thirty or forty feet away from the engine and the governor belt slips, or gets unlaced, or the pulley gets off, about the first thing the engine would do would be to jump out of the belt and by the time you get to it, it will be having a mighty lively time all alone. This might happen once and do no harm, and it might happen again and do a great deal of damage, and you are being paid to run the engine and you must stay by it. The governor is not a difficult thing to handle, but it requires your attention.

Now if I should drop the governor, you might say that I had not given you any instructions about how to regulate it to speed. I really do not know whether it is worth while to say much about it, for governors are of different designs and are necessarily differently arranged for regulating, but to help young learners I will take the Waters governors which I think the most generally used on threshing and farm engines. You will find on the upper end of the valve or governor stem two little brass nuts. The upper one is a thumb nut and is made fast to the stem. The second nut is a loose jam nut. To increase the speed of the engine loosen this jam nut and take hold of the thumb nut and turn it back slowly, watching the motion of your engine all the while. When you have obtained the speed you require, run the thumb nut down as tight as you can with your fingers. Never use a wrench on these nuts. To slow or slacken the speed, loosen the jam nut as before, except that you must run it up a few turns, then taking hold of the thumb nut, turn down slowly until you have the speed required, when you again set the thumb nut secure. In regulating the speed, be careful not to press down on the stem when turning, as this will make the engine run a little slower than it will after the pressure of your hand is removed.

If at any time your engine refuses to start with an open throttle, notice your governor stem, and you will find that it has been screwed down as far as it will go. This frequently happens with a new engine, the stem having been screwed down for its protection in transportation.

In traveling through timber with an engine, be very careful not to let any over-hanging limbs come in contact with the governor.

Now I think what I have said regarding this particular governor will enable you to handle any one you may come in contact with, as they are all very much alike in these respects. It is not my intention to take time and space to describe a governor in detail. If you will follow the instructions I have given you the governor will attend to the rest.

PART SECOND

WATER SUPPLY

If you want to be a successful engineer it is necessary to know all about the pump. I have no doubt that many who read this book, cannot tell why the old wooden pump (from which he has pumped water ever since he was tall enough to reach the handle) will pump water simply because he works the handle up and down. If you don't know this I have quite a task on my hands, for you must not attempt to run an engine until you know the principle of the pump. If you do understand the old town pump, I will not have much trouble with you, for while there is no old style wooden pump used on the engine, the same principles are used in the cross head pump. Do not imagine that a cross head pump means something to be dreaded. It is only a simple lift and force pump, driven from the cross head. That is where it gets its name and it don't mean that you are to get cross at it if it don't work, for nine times out of ten the fault will be yours. Now I am well aware that all engines do not have cross head pumps and with all respect to the builders of engines who do not use them, I am inclined to think that all standard farm engines ought to have a cross head pump, because it is the most simple and is the most economical, and if properly constructed, is the most reliable.

A cross head pump consists of a pump barrel, a plunger, one vertical check valve and two horizontal check valves, a globe valve and one stop cock, with more or less piping. We will now locate each of these parts and will then note the part that each performs in the process of feeding the boiler.

You will find all, or most pump barrels, located under the cylinder of the engine. It is placed here for several reasons. It is out of the way. It is a convenient place from which to connect it to the cross head by which it is driven. On some engines it is located on the top or at the side of the cylinder and will work equally well. The plunger is connected with the cross head and in direct line with the pump barrel, and plays back and forth in the barrel. The vertical check valve is placed between the pump and the water supply. It is not absolutely necessary that the first check be a vertical one, but a check of some kind must be so placed. As the water is lifted up to the boiler it is more convenient to use a vertical check at this point. Just ahead and a few inches from the pump barrel is a horizontal check valve. Following the course of the water toward the point where it enters the boiler, you will find another check valve. This is called a "hot water check." just below this check, or between it and where the water enters the boiler, you will find a stop cock or it may be a globe valve. They both answer the same purpose. I will tell you further on why a stop cock is preferable to a globe valve. While the cross head pumps may differ as to location and arrangement, you will find that they all require the parts described and that the checks are so placed that they bear the same relation to each other. No fewer parts can be used in a pump required to lift water and force it against steam pressure. More check valves may be used, but it would not do to use less. Each has its work to do, and the failure of one defeats all the others. The pump barrel is a hollow cylinder, the chamber being large enough to admit the plunger which varies in size from 5/8 of an inch to I inch in diameter, depending upon the size of the boiler to be supplied. The barrel is usually a few inches longer than the stroke of the engine, and is provided at the cross head end with a stuffing box and nut. At the discharge end it is tapped out to admit of piping to conduct water from the pump. At the same end and at the extreme end of the travel of the plunger it is tapped for a second pipe through which the water from the supply reaches the pump barrel. The plunger is usually made of steel and turned down to fit snug in the chamber, and is long enough to play the full stroke of engine between the stuffing box and point of supply and to connect with the driver on the cross head. Now, we will take it for granted, that, to begin with, the pump is in good order, and we will start it up stroke at a time and watch its work. Now, if everything be in good order, we should have good water and a good hard rubber suction hose attached to the supply pipe just under the globe valve. When we start the pump we must open the little pet cock between the two horizontal check valves. The globe valve must be open so as to let the water in. A check valve, whether it is vertical or horizontal, will allow water to pass through it one way only, if it is in good working order. If the water will pass through both ways, it is of no account. Now, the engine starts on the outward stroke and draws the-plunger out of the chamber. This leaves a space in the barrel which must be filled. Air cannot get into it, because the pump is in perfect order, neither can the air get to it through the hose, as it is in the water, so that the pressure on the outside of the water causes it to flow up through the pipes through the first check valve and into the pump barrel, and fills the space, and if the engine has a I2-inch stroke, and the plunger is I inch in diameter, we have a column of water in the pump I2 inches long and I inch in diameter.

The engine has now reached its outward stroke and starts back. The plunger comes back with it and takes the space occupied by the water, which must get out of the way for the plunger. The water came up through the first check valve, but it can't get back that way as we have stated. There is another check valve just ahead, and as the plunger travels back it drives the water through this second check. When the plunger reaches the end of the backward stroke, it has driven the water all out. It then starts forward again, but the water which has been driven through the second check cannot get back and this space must again be filled from supply, and the plunger continues to force more water through the second check, taking four or five strokes of the plunger to fill the pipes between the second check valve and the hot water check valve. If the gauge shows I00 pounds of steam, the hot water check is held shut by I00 pounds pressure, and when the space between the check valves is filled with water, the next stroke of the plunger will force the water through the hot water check valve, which is held shut by the I00 pounds steam pressure so that the pump must force the water against this hot water check valve with a power greater than I00 pounds pressure. If the pump is in good condition, the plunger does its work and the water is forced through into the boiler.

A clear sharp click of the valves at each stroke of the plunger is certain evidence that the pump is working well.

The small drain cock between the horizontal checks is placed there to assist in starting the pump, to tell when the pump is working and to drain the water off to prevent freezing. When the pump is started to work and this drain cock is opened, and the hot water in the pipes drained off, the globe valve is then opened, and after a few strokes of the plunger, the water will begin to flow out through the drain cock, which is then closed, and you may be reasonably certain that the pump is working all right. If at any time you are in doubt as to whether the pump is forcing the water through the pipes, you can easily ascertain by opening this drain cock. It will always discharge cold water when the pump is working. Another way to tell if the pump is working, is by placing your hand on the first two check valves. If they are cold, the pump is working all right, but if they are warm, the cold water is not being forced through them.

A stop cock should be used next to boiler, as you ascertain whether it is open or shut by merely looking at it, while the globe valve can be closed by some meddlesome party and you would not discover it, and would burst some part of your pump by forcing water against it.

PART THIRD

It is very important when the pump fails to work to ascertain what the trouble is. If it should stop suddenly, examine the tank and ascertain if you have any water. If you have sufficient water, it may be that there is air in the pump chamber, and the only way that it can get in is through the stuffing box around the plunger, if the pipes are all tight. Give this stuffing nut a turn, and if the pump starts off all right, you have found the trouble, and it would be well to re-pack the pump the first chance you get.

If the trouble is not in the stuffing box, go to the tank and see if there is anything over the screen or strainer at the end of the hose. If there is not, take hold of the hose and you can tell if there is any suction. Then ascertain if the water flows in and then out of the hose again. You can tell this by holding your hand loosely over the end of the hose. If you find that it draws the water in and then forces it out again, the trouble is with the first check valve. There is something under it which prevents its shutting down. If, however, you find that there is no suction at the end of hose examine the second check. If there should be something under it, it would prevent the pump working, because the pump forces the water through it; and, as the plunger starts back, if the check fails to hold, the water flows back and fills the pump barrel again and there would be no suction.

The trouble may, however, be in the hot water check, and it can always be told whether it is in the second check or hot water check by opening the little drain cock. If the water which goes out through it is cold, the trouble is in the second check; but, if hot water and steam are blown out through this little drain cock, the trouble is in the hot water check, or the one next to the boiler. This check must never be tampered with without first turning the stop cock between this check and the boiler. The valve can then be taken out and the obstruction removed. Be very careful never to take out the hot water check without closing the stop cock, for if you do you will get badly scalded; and never start the pump without opening this valve, for if you do, it will burst the pump.

The obstruction under the valves is sometimes hard to find. A young man in southern Iowa got badly fooled by a little pebble about the size of a pea, which got into the pipe, and when he started his pump the pebble would be forced up under the check and let the water back. When he took the check out the pebble was not there, for it had dropped back into the pipe. You will see that it is necessary to make a careful examinations and not get mad, pick up a wrench and whack away at the check valve, bruising it so that it will not work. Remember that it would work if it could, and make up your mind to find out why, it don't work. A few years ago I was called several miles to see an engine on which the pump would not work. The engine had been idle for two days and the engineer had been trying all that time to make the pump work. I took the cap off of the horizontal check, just forward of the pump barrel, and took the valve out and discovered that the check was reversed. I told the engineer that if he would put the check in so that the water could get through, he would have no more trouble. This fellow had lost his head. He was completely rattled. He insisted that "the valve had always been on that way," although the engine had been run two years.

Now the facts in this case were as follows: The old check valve in place of the one referred to had been one known as a stem valve, or floating valve. This stem by some means, had broken off but it did not prevent the valve from working. The stem, however, worked forward till it reached the hot water check, and lodged under the valve, which prevented this check from working and his pump refused to work, the engineer soon found where the stem had broken off, and instead of looking for the stem, sent to town for a new check, after putting this on the pump now refused to work for two reasons. One was, he had not removed the broken stem from the hot water check, and another was, that the new check was in wrong end to. After wasting another hour or two he finally found and-removed the stem from the hot water check, but his pump still refused to work. And then as the boys say, "he laid down," and when I called his attention to the new valve being in wrong, he was so completely rattled that he made use of the above expression.

There are other causes that would prevent the pump working besides lack of packing and obstructions under the valves. The valve may stick. When it is raised to allow the water to flow through, it may stick in the valve chamber and refuse to settle back in the seat. This may be caused by a little rough place in the chamber, or a little projection on the valve, and can generally be remedied by tapping the under side of check with a wrench or hammer. Do not strike it so hard as to bruise the check, but simply tap it. If this don't remedy the trouble, take the valve out, bore a hole in a board about I/2 inch deep and large enough to permit the valve to be turned. Drop a little emery dust in this hole. If you haven't any emery dust, scrape some grit from a common whetstone. If you have no whetstone, put some fine sand or gritty soil in the hole, put the valve on top of it, put your brace on the valve and turn it vigorously for a few minutes, and you will remove all roughness.

Constant use may sometimes make a burr on the valve which will cause it to stick. Put it through the above course and it will be as good as new. If this little process was generally known, a great deal of trouble and annoyance could be avoided.

It will not be necessary to describe other styles of pumps. If you know how to run the cross head pump, you can run any of the others. Some engines have cross head pump only. Others have an independent pump. Others have an injector, or inspirator, and some have both cross head pump and injector. I think a farm engine should be supplied with both.

It is neither wise nor necessary to go into a detailed description of an injector. The young reader will be likely to become convinced if an injector works for five minutes, it will continue to work, if the conditions remain the same. If the water in the tank does not become heated, and no foreign substance is permitted to enter the injector, there is nothing to prevent its working properly as long as the conditions are within the range of a good injector. It is a fact that with all injectors as the vertical distance the injector lifts is increased, it requires a greater steam pressure to start the injector, and the highest steam pressure at which the injector will work is greatly decreased. If the feed water is heated, a greater steam pressure is required to start the injector and it will not work with as high steam pressure. The capacity of an injector is always decreased as the lift is increased, or the feed water heated. To obtain the most economical results the proper sized injector must be used. When the exact quantity of water consumed per hour is known it can be easily determined from the capacities given in the price lists which sized injector must be selected.

An injector must always be selected having a maximum capacity in excess of the water consumed. If the exact amount of water consumed per hour is not known, and cannot be easily determined, the proper size can be approximately determined from the nominal H. P. of the boiler. The usual custom has been to allow 7 1/2 gallons of water per hour, which is a safe rule for the ordinary type of boiler.

WHAT A GOOD INJECTOR OUGHT TO DO.

With cold feed water, a good injector with a two foot lift ought to start with 25 pounds pressure and work up to I50 pounds. With 8 foot lift, ought to start at 30 pounds and work up to I30. With feed water heated to I00 degrees Fahrenheit it should start with the same lift, that is, will say 2 foot, at 26 and work Up to I20, and at 8, from 33 up to I00. You will see by this that conditions, consisting of variation of temperature in the feed water and different lifts, change the efficiency of your injector very materially, and the water can soon get beyond the ability of your injector to work at all. The above refers more particularly to the single tube injector. The double tube injector under the same conditions as above should work from I4 pounds to 250, and from I5 to 2I0, but as this injector is not generally used on farm engines you will most likely not meet with it very often.

The injector should not be placed too near the boiler, as the heat from it will make it difficult to start the injector each time after it has been standing idle.

If the injector is so hot that it will not lift the cold water, there is no way of cooling it except by applying the water on the outside. This is most effectively done by covering the injector with a cloth and pouring water over the cloth. If, after the injector has become cool, it still refuses to work, you may be sure that there is some obstruction in it that must be removed. This can be done by taking off the cap, or plug-nut, and running a fine wire through the cone valve or cylinder valve. The automatic injector requires only the manipulation of the steam valve to start it. There are other makes that require, first: that the injector be given steam and then the water. To start an injector requires some little tact, (and you will discover that tact is the handiest tools you can have to make you a good engineer). To start an injector of the Pemberthy type; first give it sufficient steam to lift the water, allowing the water to escape at overflow for a moment or long enough to cool the injector, then with a quick turn shut off and open up the supply which requires merely a twist of the wrist.

If the injector fails to take hold at once don't get ruffled but repeat the above move a few times and you will soon start it, and if you have tact, (it is only another word for natural ability) you will need no further instructions to start your injector. But remember that no injector can work coal cinders or chaf and that all joints must be air tight. Don't forget this.

It is now time to give some attention to the heater. While the heater is no part of the pump, it is connected with it and does its work between the two horizontal check valves. Its purpose is to heat the water before it passes into the boiler. The water on its way from the pump to the boiler is forced through a coil of pipes around which the exhaust steam passes on its way from the cylinder to the exhaust nozzle in the smokestack.

The heaters are made in several different designs, but it is not necessary to describe all of them, as they require little attention and they all answer the same purpose. The most of them are made by the use of a hollow bedplate with steam fitted heads or plates. The water pipe passes through the plate at the end of the heater into the hollow chamber, and a coil of pipes is formed, and the pipe then passes back through the head or plate to the hot water check valve and into the boiler.

The steam enters the cylinder from the boiler, varying in degrees of heat from 300 to 500. After acting on the piston head, it is exhausted directly into the chamber or hollow bed-plate through which the pipes pass. The water, when it enters the heater, is as cold as when it left the tank, but the steam which surrounds the pipes has lost but little of its heat, and by the time the water passes through the coil of pipes it is heated to nearly boiling point and can be introduced into the boiler with little tendency to reduce the steam. This use of the exhaust steam is economical, as it saves fuel, and it would be injurious to pump cold water directly into a hot boiler.

If your engine is fitted with both cross head pump and injector, you use the injector for pumping water when the engine is not running. The injector heats the water almost as hot as the heater. If your engine is running and doing no work, use your injector and stop the pump, for, while the engine is running light, the small amount of exhaust steam is not sufficient to heat the water and the pressure will be reduced rapidly. You will understand, therefore, that the injector is intended principally for an emergency rather than for general use. It should always be kept in order, for, should the pump refuse to work, you have only to start your injector and use it until such time as you can remedy the trouble.

We have now explained how you get your water supply. You understand that you must have water first and then fire. Be sure that you have the water supply first.

THE BLOWER

The blower is an appliance for creating artificial draught and consists of a small pipe leading from some point above the water line into the smoke stack, directly over the tubes, and should extend to the center of stack and terminate with a nozzle pointing directly to top and center of stack; this pipe is fitted with a globe valve. When it is required to rush your fire, you can do so by opening this globe and allowing the steam to escape into the stack. The force of the steam tends to drive the air out of the stack and the smoke box, this creates a strong draught. But you say, "What if I have no steam?" Well, then don't blow, and be patient till you have enough to create a draught; and it has been my experience that there is nothing gained by putting on the blower before having fifteen pounds of steam, as less pressure than this will create but little draught and the steam will escape about as fast as it is being generated. Be patient and don't be everlastingly punching at the fire. Get your fuel in good shape in fire box and shut the door and go about your business and let the fire burn.

Must the blower be used while working the engine. No. The exhaust steam which escapes into the stack, does exactly what we stated the blower does, and if it is necessary to use the blower in order to keep up steam, you can conclude that your engine is in bad shape, and yet there are times when the blower is necessary, even when your engine is in the best of condition. For instance, when you have poor fuel and are working your engine very light, the exhaust steam may not be sufficient to create enough draught for poor coal, or wet or green wood. But if you are working your engine hard the blower should never be used; if you have bad fuel and it is necessary to stop your engine you will find it very convenient to put on the blower slightly, in order to hold your steam and keep the fire lively until you start again.

It will be a good plan for you to take a look at the nozzle on blower now and then, to see that it does not become limed up and to see that it is not turned to the side so that it directs the steam to the side of stack. Should it do this, you will be using the steam and getting but little, if any, benefit. It will also be well for you to remember that you can create too much draught as well as too little; too much draught will consume your fuel and produce but little steam.

A GOOD FIREMAN.

What constitutes a good fireman? You no doubt have heard this expression: "Where there is so much smoke, there must be some fire." Well, that is true, but a good fireman don't make much smoke. We are speaking of firing with coal, now. If I can see the smoke ten miles from a threshing engine, I can tell what kind of a fireman is running the engine; and if there is a continuous cloud of black smoke being thrown out of the smokestack, I make up my mind that the engineer is having all he can do to keep the steam up, and also conclude that there will not be much coal left by the time he gets through with the job; while on the other hand, should I see at regular intervals a cloud of smoke going up, and lasting for a few moments, and for the next few moments see nothing, then I conclude that the engineer of that engine knows his business, and that he is not working hard; he has plenty of steam all the time, and has coal left when he is through. So let us go and see what makes this difference and learn a valuable lesson. We will first go to the engine that is making such a smoke, and we will find that the engineer has a big coal shovel just small enough to allow it to enter the fire door. You will see the engineer throw in about two, or perhaps three shovels of coal and as a matter of course, we will see a volume of black smoke issuing from the stack; the engineer stands leaning on his shovel watching the steam gauge, and he finds that the steam don't run up very fast, and about the time the coal gets hot enough to consume the smoke, we will see him drop his shovel, pick up a poker, throw open the fire door and commence a vigorous punching and digging at the fire. This starts the black smoke again, and about this time we will see him down on his knees with his poker, punching at the underside of the grate bars, about the time he is through with this operation the smoke is coming out less dense, and he thinks it time to throw in more coal, and he does it. Now this is kept up all day, and you must not read this and say it is overdrawn, for it is not, and you can see it every day, and the engineer that fires in this way, works hard, burns a great amount of coal, and is afraid all the time that the steam will run down on him.

Before leaving him let us take a look at his firebox, and we will see that it is full of coal, at least up to the level of the door. We will also see quite a pile of ashes under the ash pan. You can better understand the disadvantage of this way of firing after we visit the next man. I think a good way to know how to do a thing, is to know also, how not to do it.

Well, we will now go across to the man who is making but little smoke, and making that at regular intervals. We will be likely to find that he has only a little hand shovel. He picks this up, takes up a small amount of coal, opens the fire door and spreads the coal nicely over the grates; does this quickly and shuts the door; for a minute black smoke is thrown out, but only for a minute. Why? Because he only threw in enough to replenish the fire, and not to choke it in the least, and in a minute the heat is great enough to consume all the smoke before it reaches the stack, and as smoke is unconsumed fuel, he gains that much if he can consume it. We will see this engineer standing around for the next few minutes perfectly, at ease. He is not in the least afraid of his steam going down. At the end of three to five minutes, owing to the amount of work he is doing, you will see him pick up his little shovel and throw in a little coal; he does exactly as he did before, and if we stay there for an hour we will not see him pick up a poker. We will look in at his firebox, and we will see what is called a "thin fire," but every part of the firebox is hot. We will see but a small pile of ashes under the engine and he is not working hard.

If you happen to be thinking of buying an engine, you will say that this last fellow "has a dandy engine." "That is the kind of an engine I want," when the facts in the case may be that the first man may have a better engine, but don't know how to fire it. Now, don't you see how important it is that you know how to fire an engine? I am aware that some big coal wasters will say, "It is easy to talk about firing with a little hand shovel, but just get out in the field as we do and get some of the kind of fuel we have to burn, and see how you get along." Well, I am aware that you will have some bad coal. It is much better to handle bad coal in a good way than to handle good coal in a bad way. Learn to handle your fuel in the proper way and you will be a good fireman. Don't get careless and then blame the coal for what is your own fault. Be careful about this, you might give yourself away. I have seen engineers make a big kick about the fuel and claim that it was no good, when some other fellow would take hold of the engine and have no trouble whatever. Now, this is what I call a clean give away on the kicker.

Don't allow any one to be a better fireman than yourself. You will see a good fireman do exactly as I have stated. He fires often, always keeps a level fire, never allows the coal to get up to the lower tubes, always puts in coal before the steam begins to drop, keeps the fire door open as little as possible, preventing any cold air from striking the tubes, which will not only check the steam, but is injurious to the boiler.

It is no small matter to know just how to handle your dampers; don't allow too much of an opening here. You will keep a much more even fire by keeping the damper down, just allowing draught enough to allow free combustion; more than this is a waste of heat.

Get all out of the coal you can, and save all you get. Learn the little points that half the engineers never think of.

WOOD

You will find wood quite different in some respects, but the good points you have learned will be useful now. Fire quick and often, but unlike coal, you must keep your fire box full. Place your wood as loosely as possible. I mean by this, place in all directions to allow the draft to pass freely through it. Keep adding a couple sticks as fast as there is room for it; don't disturb the under sticks. Use short wood and fire close to the door. When firing with wood I would advise you to keep your screen down. There is much more danger of setting fire with wood than with coal.

If you are in a dangerous place, owing to the wind and the surroundings, don't hesitate to state your fears to the man for whom you are threshing. He is not supposed to know the danger as well as you, and if, after your advice, he says go ahead, you have placed the responsibility on him; but even after you have done this, it sometimes shows a good head to refuse to fire with wood, especially when you are required to fire with old rails, which is a common fuel in a timbered country. While they make a hot fire in a firebox, they sometimes start a hot one outside of it. It is part of your business to be as careful as you can. What I mean is take reasonable precaution, in looking after the screen in stack. If it burns out get a new one. With reasonable diligence and care, you will never set anything on fire, while on the other hand, a careless engineer may do quite a lot of damage.

There is fire about an engine, and you are provided with the proper appliances to control it. See that you do it.

WHY GRATES BURN OUT

Grates burn through carelessness. You may as well make up your mind to this at the start. You never saw grate bars burn out with a clean ash box. They can only be burned by allowing the ashes to accumulate under them till they exclude the air when the bars at once become red hot. The first thing, they do is to warp, and if the ashes are not removed at once, the grate bar will burn off. Carelessness is neglecting something which is a part of your business, and as part of it is to keep your ash box clean, it certainly is carelessness if you neglect it. Your coal may melt and run down on the bars, but if the cold air can get to the grates, the only damage this will do is to form a clinker on the top of grates, and shut off your draught. When you find that you have this kind of coal you will want to look after these clinkers.

Now if you should have good success in keeping steam, keep improving on what you know, and if you run on 1000 pounds of coal today, try and do it with 900 tomorrow. That is the kind of stuff a good fireman is made of.

But don't conclude that you can do the same amount of work each day in the week on the same amount of fuel, even should it be of the same kind. You will that with all your care and skill, your engine will differ very materially both as to the amount of fuel and water that it will require, though the conditions may apparently be the same.

This may be as good a time as any to say to you, remember that a blast of cold air against the tubes is a bad thing, so be careful about your firedoor; open it as little as, possible; when you want to throw in fuel, don't open the door, and then go a rod away after a shovel of coal; and I will say here that I have seen this thing done by men who flattered themselves that they were about at the top in the matter of running an engine. That kind of treatment will ruin the best boiler in existence. I don't mean that once or twice will do it, but to keep it up will do it. Get your shovel of coal and when you are ready to throw it in, open the door quickly and close it at once. Make it one of your habits to do this, and you will never think of doing it in any other way. If it becomes necessary to stop your engine with a hot fire and a high pressure of steam, don't throw your door open, but drop your damper and open the smoke box door.

If, however, you only expect to stop a minute or two, drop your damper, and start your injector if you have one. If you have none, get one.

An independent boiler feeder is a very nice thing, if constructed on the proper principles. You can't have your boiler too well equipped in this particular.

PART FOUR.

A boiler should be kept clean, outside and inside. Outside for your own credit, and inside for the credit of the manufacturers. A dirty boiler requires hard firing, takes lots of fuel, and is unsatisfactory in every way. The best way to keep it clean is not to let it get dirty. The place to begin work, is with your "water boy," pursuade him to be very careful of the water he brings you, if you can't succeed in this, ask him to resign.

I have seen a water-hauler back into a stream, and then dip the water from the lowerside of tank, the muddy water always goes down stream and the wheels stir up the mud; and your bright water hauler dips it into the tank. While if he had dipped it from the upper side he would have gotten clear water. However, the days of dipping water are past, but a water boy that will do as I have stated is just as liable to throw his hose into the muddy water or lower side of tank as on the upper side, where it is clear. See that he keeps his tank clean. We have seen tanks with one-half inch of mud in the bottom. We know that there are times when you are compelled to use muddy water, but as soon as it is possible to get clear water make him wash out his tank, and don't let him haul it around till the boiler gets it all.

Allow me just here to tell you how to construct a good tank for a traction engine. You can make the dimensions to suit yourself, but across the front end and about two feet back fit a partition or second head; in the center of this head and about an inch from the bottom bore a two inch hole. Place a screen over this hole on the side next the rear, and on the other side, or side next front end, put a valve. You can construct the valve in this way: Take a piece of thick leather, about four inches long, and two and a half inches wide; fit a block of wood (a large bung answers the purpose nicely) on one end, trimming the leather around one side of the wood, then nail the long part of the valve just above the hole, so that the valve will fit nicely over the hole in partition. When properly constructed, this valve will allow the water to flow into the front end of tank, but will prevent its running back. So, when you are on the road with part of a tank of water, and start down hill, this front part fills full of water, and when you start up hill, it can not get back, and your pumps will work as well as if you had a full tank of water, without this arrangement you cannot get your pumps to work well in going up a steep hill with anything less than a full tank. Now, this may be considered a little out of the engineer's duty, but it will save lots of annoyance if he has his tank supplied with this little appliance, which is simple but does the business.

A boiler should be washed out and not blown out, I believe I am safe in saying that more than half the engineers of threshing engines today depend on the "blowing out" process to clean their boilers. I don't intend to tell you to do anything without giving my reasons. We will take a hot boiler, for instance; say, 50 pounds steam. We will, of course, take out the fire. It is not supposed that anyone will attempt to blow out the water with any fire in the firebox. We will, after removing the fire, open the blow-off valve, which will be found at the bottom or lowest water point. The water is forced out very rapidly with this pressure, and the last thing that comes out is the steam. This steam keeps the entire boiler hot till everything is blown out, and the result is that all the dirt, sediment and lime is baked solid on the tubes and side of firebox. But you say you know enough to not blow off at 50 pounds pressure. Well, we will say 5 pounds, then. You will admit that the boiler is not cold by any means, even at only 5 pounds, and if you know enough not to blow off at 50 pounds, you certainly know that at 5 pounds pressure the damage is not entirely avoided. As long as the iron is hot, the dirt will dry out quickly, and by the time the boiler is cold enough to force cold water through it safely, the mud is dry and adheres closely to the iron. Some of the foreign matter will be blown out, but you will find it a difficult matter to wash out what sticks to the hot iron.

I am aware that some engineers claim that the boiler should be blown out at about 5 pounds or 10 pounds pressure, but I believe in taking the common sense view. They will advise you to blow out at a low pressure, and then, as soon as the boiler is cool enough, to wash it thoroughly.

Now, if you must wait till the boiler is cool before washing, why not let it cool with the water in it? Then, when you let the water out, your work is easy, and the moment you begin to force water through it, you will see the dirty water flowing out at the man or hand hole. The dirt is soft and washes very easily; but, if it had dried on the inside of boiler while you were waiting for it to cool, you would find it very difficult to wash off. .

You say I said to force the water through the boiler, and to do this you must use a force pump. No engineer ought to attempt to run an engine without a force pump. It is one of the necessities. You say, can't you wash out a boiler without a force pump? Oh, yes! You can do it just like some people do business. But I started out to tell you how to keep your boiler clean, and the way to do it is to wash it out, and the way to wash it out is with a good force pump. There are a number of good pumps made, especially for threshing engines. They are fitted to the tank for lifting water for filling, and are fitted with a discharge hose and nozzle.

You will find at the bottom of boiler one or two hand hole plates-if your boiler has a water bottom-if not, they will be found at the bottom of sides of firebox. Take out these hand hole plates. You will also find another plate near the top, on firebox end of boiler; take this out, then open up smoke box door and you will find another hand hole plate or plug near lower row of tubes; take this out, and you are ready for your water works, and you want to use them vigorously; don't throw in a few buckets of water, but continue to direct the nozzle to every part of the boiler, and don't stop as long as there is any muddy water flowing at the bottom hand holes. This is the way to clean your boiler, and don't think that you can be a success as an engineer without this process, and once a week is none too often. If you want satisfactory results from your engine, you must keep a clean boiler, and to keep it clean requires care and labor. If you neglect it you can expect trouble. If you blow out your boiler hot, or if the mud and slush bakes on the tubes, there is soon a scale formed on the tubes, which decreases the boiler's evaporating capacity. You, therefore, in order to make sufficient amount of steam, must increase the amount of fuel, which of itself is a source of expense, to say nothing of extra labor and the danger of causing the tubes to leak from the increased heat you must produce in the firebox in order to make steam sufficient to do the work.

You must not expect economy of fuel, and keep a dirty boiler, and don't condemn a boiler because of hard firing until you know it is clean, and don't say it is clean when it can be shown to be half full of mud.

SCALE

Advertisements say that certain compounds will prevent scale on boilers, and I think they tell the truth, as far as they go; but they don't say what the result may be on iron. I will not advise the use of any of these preparations, for several reasons. In the first place, certain chemicals will successfully remove the scale formed by water charged with bicarbonate of lime, and have no effect on water charged with sulphate of lime. Some kinds of bark-summac, logwood, etc.,-are sufficient to remove the scale from water charged with magnesia or carbonate of lime, but they are injurious to the iron owing to the tannic acid with which they are charged. Vinegar, rotten apples, slop, etc., owing to their containing acetic acid, will remove scale, but this is even more injurious to the iron than the barks. Alkalies of any kind, such as soda, will be found good in water containing sulphate of lime, by converting it into a carbonate and thereby forming a soft scale, which is easily washed out; but these have their objections, for, when used to excess, they cause foaming.

Petroleum is not a bad thing in water where sulphate of lime prevails; but you should use only the refined, as crude oil sometimes helps to form a very injurious scale. Carbonate of soda and corn-starch have been recommended as a scale preventative, and I am inclined to think they are as good as anything, but as we are out in the country most of the time I can tell you of a simple little thing that will answer the same purpose, and can usually be had with little trouble. Every Monday morning just dump a hatful of potatoes into your boiler, and Saturday night wash the boiler out, as I have already suggested, and when the fall's run is over there will not be much scale in the boiler.

CLEAN FLUES.

We have been urging you to keep your boiler clean. Now, to get the best results from your fuel, it will also be necessary to keep your flues clean; as soot and ashes are non-conductors of heat, you will find it very difficult to get up steam with a coating of soot in your tubes. Most factories furnish with each engine a flue cleaner and rod. This cleaner should be made to fit the tubes snug, and should be forced through each separate tube every morning before building a fire. Some engineers never touch their flues with a cleaner, but when they choke the exhaust sufficiently to create such a draught as to clean the flues, they are working the engine at a great disadvantage, besides being much more liable to pull the fire out at the top of smokestack. If it were not necessary to create draught by reducing your exhaust nozzle, your engine would run much nicer and be much more powerful if your nozzle was not reduced at all. However, you must reduce it sufficiently to give draught, but don't impair the power by making the engine clean its own flues. I think ninety per cent of the fires started by. traction engines can be traced to the engineer having his engine choked at the exhaust nozzle. This is dangerous for the reason that the excessive draught created throws fire out at the stack. It cuts the power of the engine by creating back pressure. We will illustrate this: Suppose you close the exhaust entirely, and the engine would not turn itself. If this is true, you can readily understand that partly closing it will weaken it to a certain extent. So, remember that the nozzle has something to do with the power of the engine, and you can see why the fellow that makes his engine clean its own flues is not the brightest engineer in the world.

While it is not my intention to encourage the foolish habit of pulling engines, to see which is the best puller, should you get into this kind of a test, you will show the other fellow a trick by dropping the exhaust nozzle off entirely, and no one need know it. Your engine will not appear to be making any effort, either, in making the pull. Many a test has been won more through the shrewdness of the operator than the superiority of the engine.

The knowing of this little trick may also help you out of a bad hole some time when you want a little extra power. And this brings us to the point to which I want you to pay special attention. The majority of engineers, when they want a little extra power, give the safety valve a twist.

Now, I have already told you to carry a good head of steam, anywhere from 100 to 120 pounds of steam is good pressure and is plenty, and if you have your valve set to blow off at 115, let it be there; and don't screw it down every time you want more power, for if you do you will soon have it up to I25, and should you want more steam at some other time you will find yourself screwing it down again, and what was really intended for a safety valve loses all its virtue as a safety, as far as you and those around you are concerned. If you know you have a good boiler you are safe in setting it at 125 pounds, provided you are determined to not set it up to any higher pressure. But my advice to you is that if your engine won't do the work required of it at 115 pounds, you had best do what you can with it until you can get a larger one.

A safety valve is exactly what its name implies, and there should be a heavy penalty for anyone taking that power away from it.

If you refuse to set your safety down at any time, it does not imply that you are afraid of your boiler, but rather you understand your business and realize your responsibility.

I stated before what you should do with the safety valve in starting a new engine. You should also attend to this part of it every few days. See that it does not become slow to work. You should note the pressure every time it blows off; you know where it ought to blow off, so don't allow it to stick or hold the steam beyond this pressure. If you are careful about this, there is no danger about it sticking some time when you don't happen to be watching the gauge. The steam gauge will tell you when the pop ought to blow off, and you want to see that it does it.

PART FIVE

STEAM GAUGE

Some engineers call a steam gauge a "clock." I suppose they do this because they think it tells them when it is time to throw in coal, and when it is time to quit, and when it is time for the safety valve to blow off. If that is what they think a steam gauge is for, I can tell them that it is time for them to learn differently.

It is true that in a certain sense it does tell the engineer when to do certain things, but not as a clock would tell the time of day. The office of a steam gauge is to enable you to read the pressure on your boiler at all times, the same as a scale will enable you to determine the weight of any object.

As this is the duty of the steam gauge, it is necessary that it be absolutely correct. By the use of an unreliable gauge you may become thoroughly bewildered, and in reality know nothing of what pressure you are carrying.

This will occur in about this way: Your steam gauge becomes weak, and if your safety is set at 100 pounds, it will show 100 or even more before the pop allows the steam to escape; or if the gauge becomes clogged, the pop may blow off when the gauge only shows go pounds or less. This latter is really more dangerous than the former. As you would most naturally conclude that your safety was getting weak, and about the first thing you would do would be to screw it down so that the gauge would show 100 before the pop would blow off, when in fact you would have 100 or more.

So you can see at once how important it is that your gauge and safety should work exactly together, and there is but one way to make certain of this, and that is to test your steam gauge. If you know the steam gauge is correct, you can make your safety valve agree with it; but never try to make it do it till you know the gauge is reliable.

HOW TO TEST A STEAM GAUGE

Take it off, and take it to some shop where there is a steam boiler in active use; have the engineer attach your gauge where it will receive the direct pressure, and if it shows the same as his gauge, it is reasonable to suppose that your gauge is correct. If the engineer to whom you take your gauge should say he thinks his gauge is weak, or a little strong, then go somewhere else. I have already told you that I did not want you to think anything about your engine-I want you to know it. However, should you find that your gauge shows when tested with another gauge, that it is weak, or unreliable in any way, you want to repair it at once, and the safest way is to get a new one; and yet I would advise you first to examine it and see if you cannot discover the trouble. It frequently happens that the pointer becomes loosened on the journal or spindle, which attaches it to the mechanism that operates it. If this is the trouble, it is easily remedied, but should the trouble prove to be in the spring, or the delicate mechanism, it would be much more satisfactory to get a new one.

In selecting a new gauge you will be better satisfied with a gauge having a double spring or tube, as they are less liable to freeze or become strained from a high pressure, and the double spring will not allow the needle or pointer to vibrate when subject to a shock or sudden increase of pressure, as with the single spring. A careful engineer will have nothing to do with a defective steam gauge or an unreliable safety valve. Some steam gauges are provided with a seal, and as long as this seal is not broken the factory will make it good.

FUSIBLE PLUG

We have told you about a safety valve, we will now have something to say of a safety plug. A safety, or fusible plug, is a hollow brass plug or bolt, screwed into the top crown sheet. The hole through the plug being filled with some soft metal that will fuse at a much less temperature than is required to burn iron. The heat from the firebox will have no effect on this fusible plug as long as the crown sheet is covered with water, but the moment that the water level falls below the top of the crown sheet, thereby exposing the plug, this soft metal is melted and runs out, allows the steam to rush down through the opening in the lug, putting out the fire and preventing any injury to the boiler. This all sounds very nice, but I am free to confess that I am not an advocate of a fusible plug. After telling you to never allow the water to get low, and then to say there is something to even make this allowable, sounds very much like the preacher who told his boy "never to go fishing on Sunday, but if he did go, to be sure and bring home the fish." I would have no objection to the safety plug if the engineer did not know it was there. I am aware that some states require that all engines be fitted with a fusible plug. I do not question their good intentions, but I do question their good judgment. It seems to me the are granting a license to carelessness. For instance, an engineer is running with a low gauge of water, owing possibly to the tank being delayed longer than usual, he knows the water is getting low, but he says to himself, "well, if the water gets too low I will only blow out the plug," and so he continues to run until the tank arrives. If the plug holds, he at once begins to pump in cold water, and most likely does it on a very hot sheet, which of itself, is something he never should do; and if the plug does blow out he is delayed a couple of hours, at least, before he can put in a new plug and get up steam again. Now suppose he had not had a soft plug (as they are sometimes called). He would have stopped before he had low water. He would not even have had a hot crown sheet, and would only have lost the time he waited on the tank. This is not a fancied circumstance by any means, for it happens every day. The engineer running an engine with a safety plug seldom stops for a load of water until he blows out the plug. It frequently happens that a fusible plug becomes corroded to such an extent that it will stand a heat sufficient to burn the iron. This is my greatest objection to it. The engineer continues to rely on it for safety, the same as if it were in perfect order, and the ultimate result is he burns or cracks his crown sheet. I have already stated that I have no objection to the plug, if the engineer did not know it was there, so if you must use one, attend to it, and every time you clean your boiler scrape the upper or water end of the plug with a knife, and be careful to remove any corrosive matter that may have collected on it, and then treat your boiler exactly as though there was no such a thing as a safety plug in it. A safety plug was not designed to let you run with any lower gauge of water. It is placed there to prevent injury to the boiler, in case of an accident or when, by some means, you might be deceived in your gauge of water, or if by mistake, a fire was started without any water in the boiler.

Should the plug melt out, it is necessary to replace it at once, or as soon as the heat will permit you to do so. It might be a saving of time to have an extra plug always ready, then all you have to do is to remove the melted one by unscrewing it from the crown sheet and screwing the extra one in. But if you have no extra plug you must remove the first one and refill it with babbitt. You can do this by filling one end of the plug with wet clay and pouring the metal into the other end, and then pounding it down smooth to prevent any leaking. This done, you can screw the plug back into its place.

If you should have two plugs, as soon as you have melted out one replace it with the new one, and refill the other at your earliest convenience. By the time you have replaced a fusible plug a few times in a hot boiler you will conclude it is better to keep water over your crown sheet.

LEAKY FLUES

What makes flues leak? I asked this question once, and the answer was that the flues were not large enough to fill up the hole in flue sheet. This struck me as being funny at first, but on second thought I concluded it was about correct. Flues may leak from several causes, but usually it can be traced to the carelessness of some one. You may have noticed before this that I am inclined to blame a great many things to carelessness. Well, by the time you have run an engine a year or two you will conclude that I am not unjust in my suspicions. I do not blame engineers for everything, but I do say that they are responsible for a great many things which they endeavor to shift on to the manufacturer. If the flues in a new boiler leak, it is evident that they were slighted by the boiler-maker; but should they run a season or part of a season before leaking, then it would indicate that the boiler-maker did his duty, but the engineer did not do his. He has been building too hot a fire to begin with, or has, been letting his fire door stand open; or he may have overtaxed his boiler; or else he has been blowing out his boiler when too hot; or has at some time blown out with some fire in firebox. Now, any one of these things, repeated a few times, will make the best of them leak. You have been advised already not to do these things, and if you do them, or any one of them, I want to know what better word there is to express it than "carelessness."

There are other things that will make your flues leak. Pumping cold water into a boiler with a low gauge of water will do it, if it does nothing more serious. Pouring cold water into a hot boiler will do it. For instance, if for any reason you should blow out your boiler while in the field, and as you might be in a hurry to get to work, you would not let the iron cool, before beginning to refill. I have seen an engineer pour water into a boiler as soon as the escaping steam would admit it. The flues cannot stand such treatment, as they are thinner than the shell or flue sheet, and therefore cool much quicker, and in contracting are drawn from the flue sheet, and as a matter of course must leak. A flue, when once started to leak, seldom stops without being set up, and one leaky flue will start others, and what are you going to do about it? Are you going to send to a boiler shop and get a boilermaker to come out and fix them and pay him from forty to sixty cents an hour for doing it? I don't know but that you must the first time, but if you are going to make a business of making your flues leak, you had best learn how to do it yourself. You can do it if you are not too big to get into the fire door. You should provide yourself with a flue expander and a calking tool, with a machinist's hammer, (not too heavy). Take into the firebox with you a piece of clean waste with which you will wipe off the ends of the flues and flue sheet to remove any soot or ashes that may have collected around them. After this is done you will force the expander into the flues driving it well up, in order to bring the shoulder of expander up snug against the head of the flue. Then drive the tapering pin into the expander. By driving the pin in too far you may spread the flue sufficient to crack it or you are more liable, by expanding too hard, to spread the hole in flue sheet and thereby loosen other flues. You must be careful about this. When you think you have expanded sufficient, hit the pin a side blow in order to loosen it, and turn the expander about one-quarter of a turn, and drive it up as before; loosen up and continue to turn as before until you have made the entire circle of flues. Then remove the expander, and you are ready for your header or calking tool. It is best to expand all the flues that are leaking before beginning with the header.

The header is used by placing the gauge or guide end within the flue, and with your light hammer the flue can be calked or beaded down against the flue sheet. Be careful to use your hammer lightly, so as not to bruise the flues or sheet. When you have gone over all the expanded flues in this way, you, (if you have been careful) will not only have a good job, but will conclude that you are somewhat of an expert at it. I never saw a man go into a firebox and stop the leak but that he came out well pleased with himself. The fact that a firebox is no pleasant workshop may have had something to do with it. If your flues have been leaking badly, and you have expanded them, it would be well to test your boiler with cold water pressure to make sure that you have a good job.

How are you going to test your boiler? If you can attach to a hydrant, do so, and when you have given your boiler all the pressure you want, you can then examine your flues carefully, and should you find any seeping of water, you can use your beader lightly untill such leaks are stopped. If the waterworks will not afford you sufficient pressure, you can bring it up to the required pressure, by attaching a hydraulic pump or a good force pump.

In testing for the purpose of ascertaining if you have a good job on your flues, it is not necessary to put on any greater cold water pressure than you are in the habit of carrying. For instance, if your safety valve is set at one hundred and ten pounds, this pressure of cold water will be sufficient to test the flues.

Now, suppose you are out in the field and want to test your flues. Of course you have no hydrant to attach to, and you happen not to have a force pump, it would seem you were in bad shape to test your boiler with cold water. Well, you can do it by proceeding in this way: When you have expanded and beaded all the flues that were leaking, you will then close the throttle tight, take off the safety valve (as this is generally attached at the highest point) and fill the boiler full, as it is absolutely necessary that all the space in the boiler should be filled with cold water. Then screw the safety valve back in its place. You will then get back in the firebox with your tools and have someone place a small sheaf of wheat or oat straw under the firebox or under waist of boiler if open firebox, and set fire to it. The expansive force of the water caused by the heat from the burning straw will produce pressure desired. You should know, however, that your safety is in perfect order. When the water begins to escape at the safety valve, you can readily see if you have expanded your flues sufficiently to keep them from leaking.

This makes a very nice and steady pressure, and although the pressure is caused by heat, it is a cold water pressure, as the water is not heated beyond one or two degrees. This mode of testing, however, cannot be applied in very cold weather, as water has no expansive force five degrees above or five degrees below the freezing point.

These tests, however, are only for the purpose of trying your flues and are not intended to ascertain the efficiency or strength of your boiler. When this is required, I would advise you to get an expert to do it, as the best test for this is the hammer test, and only an expert should attempt it.

PART SIX

Any young engineer who will make use of what he has read will never get his engine into much trouble. Manufacturers of farm engines to-day make a specialty of this class of goods, as they endeavor to build them as simple and of as few parts as possible. They do this well knowing that, as a rule, they must be run by men who cannot take a course in practical engineering. If each one of the many thousands of engines that are turned out every year had to have a practical engineer to run it, it would be better to be an engineer than to own the engine; and manufacturers knowing this, they therefore make their engines as simple and with as little liability to get out of order as possible. The simplest form of an engine, however, requires of the operator a certain amount of brains and a willingness to do that which he knows should be done; and if you will follow the instructions you have already received, you can run your engine as successfully as any one can wish as long as your engine is in order, and, as I have just stated, it is not liable to get out of order, except from constant wear, and this wear will appear in the boxes, journals and valve. The brasses on wrist pin and cross-head will probably require your first and most careful attention, and of these two the wrist or crank box will require the most; and what is true of one is true of both boxes. It is, therefore, not necessary to take up both boxes in instructing you how to handle them. We will take up the box most likely to require your attention. This is the wrist box. You will find this box in two parts or halves. In a new engine you will find that these two halves do not meet on the wrist pin by at least one-eighth of an inch. They are brought up to the pin by means of a wedge-shaped key. (I am speaking now of the most common form of wrist boxes. If your engine should not have this key, it will have something which serves the same purpose.) As the brasses wear you can take up this wear by forcing the key down, which brings the two halves nearer together. You can continue to gradually take up this wear until you have brought them together. You will then see that it is necessary to do something, in order to take up any more wear, and this "something" is to take out the brasses and file about one-sixteenth of an inch off of each brass. This will allow you another eighth of an inch to take up in wear.

Now here is a nice little problem for you to solve and I want you to solve it to your own satisfaction, and when you do, you will thoroughly understand it, and to understand it is to never allow it to get you into trouble. We started out by saying that in a new engine you would most likely find about one-eighth of an inch between the brasses, and we said you would finally get these brasses, or halves together, and would have to take them out and file them. Now we have taken up one-eighth of an inch and the result is, we have lengthened our pitman just one-sixteenth of an inch; or in other words, the center of wrist pin and the center of cross-head are just one-sixteenth of an inch further apart than they were before any wear had taken place, and the piston head has one-sixteenth of an inch more clearance at one end, and one-sixteenth of an inch less at the other end than it had before. Now if we take out the boxes and file them so we have, another eighth of an inch, by the time we have taken up this wear, we will then have this distance doubled, and we will soon have the piston head striking the end of the cylinder, and besides, the engine will not run as smooth as it did. Half of the wear comes off of each half, and the half next to the key is brought up to the wrist pin because of the tapering key, while the outside half remains in one place. You must therefore place back of this half a thin piece of sheet copper, or a piece of tin will do. Now suppose our boxes had one-eighth of an inch for wear. When we have taken up this much we must put in one-sixteenth of an inch backing (as it is called), for we have reduced the outside half by just that amount. We have also reduced the front half the same, but as we have said, the tapering key brings this half up to its place.

Now we think we have made this clear enough and we will leave this and go back to the key again. You must remember that we stated that the key was tapering or a wedged shape, and as a wedge, is equally as powerful as a screw, and you must bear in mind that a slight tap will bring these two boxes up tight against the wrist pin. Young engineers experience more trouble with this box than with any other part of the engine, and all because they do not know how to manage it. You should be very careful not to get your box too tight, and don't imagine that every time there is a little knock about your engine that you can stop it by driving the key down a little more. This is a great mistake that many, and even old engineers make. I at one time seen a wrist pin and boxes ruined by the engineer trying to stop a knock that came from a loose fly-wheel. It is a fact, and one that has never been satisfactorily explained, that a knock coming from almost any part of an engine will appear to be in the wrist. So bear this in mind and don't allow yourself to be deceived in this way, and never try to stop a knock until you have first located the trouble beyond a doubt.

When it becomes necessary to key up your brasses, you will find it a good safe way to loosen up the set screw which holds the key, then drive it down till you are satisfied you have it tight. Then drive it back again and then with your fist drive the key down as far as you can. You may consider this a peculiar kind of a hammer, but your boxes will rarely ever heat after being keyed in this manner.

KNOCK IN ENGINES

What makes an engine knock or pound? A loose pillow block box is a good "knocker." The pillow block is a box next crank or disc wheel. This box is usually fitted with set bolts and jam nuts. You must also be careful not to set this up too tight, remembering always that a box when too tight begins to heat and this expands the journal, causing greater friction. A slight turn of a set bolt one way or the other may be sufficient to cool a box that may be running hot, or to heat one that may be running cool. A hot box from neglect of oiling can be cooled by supplying oil, provided it has not already commenced to cut. If it shows any sign of cutting, the only safe way is to remove the box and clean it thoroughly.

Loose eccentric yokes will make a knock in an engine, and it may appear to be in the wrist. You will find packing between the two halves of the yoke. Take out a thin sheet of this packing, but don't take out too much, as you are liable then to get them too tight and they may stick and cause your eccentrics to slip. We will have more to say about the slipping of the eccentrics.

The piston rod loose in cross-head will make a knock, which also appears in the wrist, but it is not there. Tighten the piston and you will stop it. The piston rod may be keyed in cross head, or it may be held in place by a nut. The key is less liable to get loose, but should it work loose a few times it may be necessary to replace it with a new one. And this is one of the things that cause a bad break when it works out or gets loose. If it gets loose it may not come out, but it will not stand the strain very long in this condition, and will break, allowing the piston to come out of cross head, and you are certain to knock out one cylinder head and possibly both of them. The nut will do the same thing if allowed to come off. So this is one of the connections that will claim your attention once in a while, but if you train your ear to detect any unusual noise you will discover it as soon as it gives the least in either key or nut.

The cross-head loose in the guides will make it knock. If the cross-head is not provided for taking up this wear, you can take off the guides and file them enough to allow them to come up to the cross-head, but it is much better to have them planed off, which insures the guides coming up square against the cross-head and thus prevent any heating or cutting.

A loose fly-wheel will most likely puzzle you more than anything else to find the knock. So remember this. The wheel may apparently be tight, but should the key be the least bit narrow for the groove in shaft, it will make your engine bump very similar to that caused by too much or too little "lead."

LEAD

What is lead? Lead is space or opening of port on steam end of cylinder, when engine is on dead center. (Dead center is the two points of disc or crank wheel at which the crank pin is in direct line with piston and at which no amount of steam will start the engine.) Different makes of engines differ to such an extent that it is impossible to give any rule or any definite amount of lead for an engine. For instance, an engine with a port six inches long and one-half inch wide would require much less lead than one with a port four inches long and one inch wide. Suppose I should say one-sixteenth of an inch was the proper lead. In one engine you would have an opening one-sixteenth of an inch wide and six inches long and in the other you would have one-sixteenth of an inch wide and four inches long; so you can readily see that it is impossible to give the amount of lead for an engine without knowing the piston area, length of port, speed, etc. Lead allows live steam to enter the cylinder just ahead of the piston at the point of finishing the stroke, and forms a "cushion," and enables the engine to pass the center without a jar. Too much lead is a source of weakness to an engine, as it allows the steam to enter the cylinder too soon and forms a back pressure and tends to prevent the engine from passing the center. It will, therefore, make your engine bump, and make it very difficult to hold the packing in stuffing box.

Insufficient lead will not allow enough steam to enter the cylinder ahead of piston to afford cushion enough to stop the inertia, and the result will be that your engine will pound on the wrist pin. You most likely have concluded by this time that "lead" is no small factor in the smooth running of an engine, and you, as a matter of course, will want to know how you are to obtain the proper lead. Well don't worry yourself. Your engine is not going to have too much lead today and not enough tomorrow. If your engine was properly set up in the first place the lead will be all right, and continue to afford the proper lead as long as the valve has not been disturbed from its original position; and this brings us to the most important duty of an engineer as far as the engine is concerned, viz: Setting the Valve.

SETTING A VALVE.

The proper and accurate setting of a valve on a steam engine is one of the most important duties that you will have to perform, as it requires a nicety of calculation and a mechanical accuracy. And when we remember also, that this is another one of the things for which no uniform rule can be adopted, owing to the many circumstances which go to make an engine so different under different conditions, we find it very difficult to give you the light on this part of your duty which we would wish to. We, however, hope to make it so clear to you that by the aid of the engine before you, you can readily understand the conditions and principles which control the valve in the particular engine which you may have under your management.

The power and economy of an engine depends largely on the accurate operation of its valve. It is, therefore, necessary that you know how to reset it, should it become necessary to do so.

An authority says, "Bring your engine to a dead center and then adjust your valve to the proper lead." This is all right as far as it goes, but how are you to find the dead center. I know that it is a common custom in the field to bring the engine to a center by the use of the eye. You may have a good eye, but it is not good enough to depend on for the accurate setting of a valve.

HOW TO FIND THE DEAD CENTER

First, provide yourself with a "tram." This you can do by taking a 1/4 inch iron rod, about 18 inches long, and bend about two inches of one end to a sharp angle. Then sharpen both ends to a nice sharp point. Now, fasten securely a block of hard wood somewhere near the face of the fly wheel, so that when the straight end of your tram is placed at a definite point in the block the other, or hook end, will reach the crown of fly wheel.

Be certain that the block cannot move from its place, and be careful to place the tram at exactly the same point on the block at each time you bring the tram into use. You are now ready to proceed to find the dead center, and in doing this re-member to turn the fly wheel always in the same direction. Now, turn your engine over till it nears one of the centers, but not quite to it. You will then, by the aid of a straight-edge make a clear and distinct mark across the guides and cross head. Now, go around to the fly wheel and place the straight end of the tram at same point on the block, and with the hook end make a mark across the crown or center of face of fly wheel; now turn your engine past the center and on to the point at which the line on cross head is exactly in line with the lines on guides. Now, place your tram in the same place as before, and make another mark across the crown of fly wheel. By the use of dividers find the exact center between the two marks made on fly wheel; mark this point with a center punch. Now, bring the fly wheel to the point at which the tram, when placed at its proper place on block, the hook end, or point, will touch this punch mark, and you will have one of the exact dead centers.

Now, turn the engine over till it nears the other center, and proceed exactly as before, remembering always to place the straight end of tram exactly in same place in block, and you will find both dead centers as accurately as if you had all the fine tools of an engine builder.

You are now ready to proceed with the setting of your valve, and as you have both dead centers to work from you ought to be able to do it, as you do not have to depend on your eye to find them, and by the use of the tram You turn your engine to exactly the same point every time you wish to get a center.

Now remove the cap on steam chest, bring your engine to a dead center and give your valve the necessary amount of lead on the steam end. Now, we have already stated that we could not give you the proper amount of lead for an engine. It is presumed that the maker of your engine knew the amount best adapted to this engine, and you can ascertain his idea of this by first allowing, we will say, about 1/16 of an inch. Now bring your engine to the other center, and if the lead at the other end is less than 1/16, then you must conclude that he intended to allow less than 1/16, but should it show more than this, then it is evident that he intended more than I/I16 lead; but in either case you must adjust your valve so as to divide the space, in order to secure the same lead when on either center. In the absence of any better tool to ascertain if the lead is the same, make a tapering wooden wedge of soft wood, turn the engine to a center and force the wedge in the opening made by the valve hard enough to mark the wood; then turn to the next center, and if the wedge enters the same distance, you are correct; if not, adjust till it does, and when you have it set at the proper place you had best mark it by taking a sharp cold chisel and place it so that it will cut into the hub of eccentric and in the shaft; then hit it a smart blow with a hammer. This should be done after you have set the set screws in eccentric down solid on the shaft. Then, at any time should your eccentric slip, you have only to bring it back to the chisel mark and fasten it, and you are ready to go ahead again.

This is for a plain or single eccentric engine. A double or reversible engine, however, is somewhat more difficult to handle in setting the valve. Not that the valve itself is any different from a plain engine, but from the fact that the link may confuse you, and while the link may be in position to run the engine one way you may be endeavoring to set the valve to run it the other way.

The proper way to proceed with this kind of an engine is to bring the reverse lever to a position to run the engine forward, then proceed to set your valve the same as on a plain engine. When you have it at the proper place, tighten just enough to keep from slipping, then bring your reverse lever to the reverse position and bring your engine to the center. If it shows the same lead for the reverse motion you are then ready to tighten your eccentrics securely, and they should be marked as before.

You may imagine that you will have this to do often. Well don't be scared about it. You may run an engine a long time, and never have to set a valve. I have heard these windy engineers (you have seen them), say that they had to go and set Mr. A's or Mr. B's valve, when the facts were, if they did anything, it was simply to bring the eccentrics back to their original position. They happened to know that most all engines are plainly marked at the factory, and all there was to do was to bring the eccentrics back to these marks and fasten them, and the valve was set. The slipping of the eccentrics is about the only cause for a valve working badly. You should therefore keep all grease and dirt away from these marks; keep the set screws well tightened, and notice them frequently to see that they do not slip. Should they slip a I/I6 part of an inch, a well educated ear can detect it in the exhaust. Should they slip a part of a turn as they will some times, the engine may stop instantly, or it may cut a few peculiar circles for a minute or two, but don't get excited, look to the eccentrics at once for the trouble.

Your engine may however act very queer some time, and you may find the eccentrics in their proper place. Then you must go into the steam chest for the trouble. The valves in different engines are fastened on valve rod in different ways. Some are held in place by jam nuts; a nut may have worked loose, causing lost motion on the valve. This will make your engine work badly. Other engines hold their valve by a clamp and pin. This pin may work out, and when it does, your engine will stop, very quickly to.

If you thoroughly understand the working of the steam, you can readily detect any defect in your cylinder or steam chest, by the use of your cylinder cocks. Suppose we try them once. Turn your engine on the forward center, now open the cocks and give the engine the steam pressure. If the steam blows out at the forward cock we know that we have sufficient lead. Now turn back to the back center, and give it steam again; if it blows out the same at this cock, we can conclude that our valve is in its proper position. Now reverse the engine and do the same thing; if the cocks act the same, we know we are right. Suppose the steam blows out of one cock all right, and when we bring the engine to the other center no steam escapes from this cock, then we know that something is wrong with the valve, and if the eccentrics are in their proper position the trouble must be in the steam chest, and if we open it up we will find the valve has become loosened on the rod. Again suppose we put the engine on a center, and on giving it steam, we find the steam blowing out at both cocks.

Now what is the trouble, for no engine in perfect shape will allow the steam to blow out of both cocks at the same time. It is one of two things, and it is difficult to tell. Either the cylinder rings leak and allow the steam to blow through, or else the valve is cut on the seat, and allows the steam to blow over. Either of these two causes is bad, as it not only weakens your engine, but is a great waste of fuel and water. The way to determine which of the two causes this, is to take off the cylinder head, turn engine on forward center and open throttle slightly. If the steam is seen to blow out of the port at open end of cylinder, then the trouble is in the valve, but if not, you will see it blowing through from forward end of cylinder, and the trouble is in the cylinder rings.

What is the remedy? Well, if the "rings" are the trouble, a new set will most likely remedy it should they be of the automatic or self-setting pattern, but should they be of the spring or adjusting pattern, you can take out the head and set the rings out to stop this blowing. As most all engines now are using the self-setting rings, you will most likely require a new set.

If the trouble is in the valve or steam chest, you had best take it off and have the valve seat planed down, and the valve seated to it. This is the safest and best way. Never attempt to dress a valve down, you are most certain to make a bad job of it.

And yet I don't like the idea of advising you not to do a thing that can be done, for I do like an engineer who does not run to the shop for every little trouble. However, unless you have the proper tools you had best not attempt it. The only safe way is to scrape them down, for if your valve is cut, you will find the valve seat is cut equally as bad, and they must both be scraped to a perfect fit. Provide yourself with a piece of flat steel, very hard, 3x4 inches by about 1/8 inch, with a perfect straight edge. With this scrape the valve and seat to a perfect flat surface, It will be a slower process than scraping wood with a piece of glass, but you can do it. Never use a chisel or a file on a valve.

LUBRICATING OIL

What is oil?

Oil is a coating for a journal, or in other words is a lining between bearings.

Did you ever stop long enough to ask yourself the question? I doubt it. A great many people buy something to use on their engine, because it is called oil. Now if the object in using oil is to keep a lining between the bearings, is it not reasonable that you use something that will adhere to that which it is to line or cover?

Gasoline will cover a journal for a minute or two, and oil a grade better would last a few minutes longer. Still another grade would do some better. Now if you are running your own engine, buy the best oil you can buy. You will find it very poor economy to buy cheap oil, and if you are not posted, you may pay price enough, but get a very poor article.

If you are running an engine for some one else, make it part of your contract that you are furnished with a good oil. You can not keep an engine in good shape with a cheap oil. You say "you are going to keep your engine clean and bright." Not if you must use a poor oil.

Poor oil is largely responsible for the fast going out of use of the link reverse among the makers of traction engines. While I think it very doubtful if this "reverse motion" can be equalled by any of the late devices. Its construction is such as to require the best grade of cylinder oil, and without this it is very unsatisfactory, (not because the valves of other valve-motions will do with a poorer grade of oil) but because its construction is such that as soon as the valve becomes dry it causes the link to jump and pound, and very soon requires repairing. While the construction of various other devices are such, that while the valve may be equally as dry it does not show the want of oil so clearly as the old style link. Yet as a fact I care not what the valve motion may be, it requires a good grade of oil.

You may ask "how am I to know when I am getting a good grade of oil." The best way is to ascertain a good brand of oil then use that and nothing else.

We are not selling oil, or advertising oil. However before I get through I propose to give you the name of a good brand of cylinder oil, a good engine oil as well as good articles of various attachments, which cut no small figure in the success you may have in running an engine.

It is not an uncommon thing for an engineer (I don't like to call him an engineer either) to fill his sight feed lubricator with ordinary engine oil, and then wonder why his cylinder squeaks. The reason is that this grade of oil cannot stand the heat in the cylinder or steam chest.

If you are carrying 90 pounds of steam you have about 320 degrees of heat in your cylinder, with 120 to 125 pounds you will have about 350 degrees of heat, and in order to lubricate your valve and valve-seat, and also the cylinder surface, you must use an oil, that will not only stand this heat but considerable more so that it will have some staying qualities.

Then if you are using a good quality of oil and your link or reverse begins to knock, it is because some part of it wants attention, and you must look after it. And here is where I want to insist that you teach your ear to be your guide. You ought to be able to detect the slightest sound that is unnatural to your engine. Your eyes may be deceived, but a well trained ear can not be fooled.

I was once invited by an engineer to come out and see how nice his engine was running. I went, and found that the engine itself was running very smooth, in fact almost noiseless, but he looked very much disappointed when I asked him why he was doing all his work with one end of cylinder. He asked me what I meant, and I had some difficulty in getting him to detect the difference in the exhaust of the two ends, in fact the engine was only making one exhaust to a revolution. He was one of those engineers who never discovered anything wrong until he could see it. Did you know that there are people in the world whose mental capacity can only grasp one idea at a time. That is when their minds are on any one object or principle they can not see or observe anything else. That was the case with this engineer, his mind had been thoroughly occupied in getting all the reciprocating (moving) parts perfectly adjusted, and if the exhaust had made all sorts of peculiar noises, he would not have discovered it.

The one idead man will not make a successful engineer. The good engineer can stand by and at a glance take in the entire engine, from tank to top of smoke stack.

He has the faculty of noting mentally, what he sees, and what he hears, and by combining the results of the two, he is enabled to size up the condition of the engine at a glance. This, however, only come with experience, and verges on expertness. And if you wish to be an expert, learn to be observing.

It is getting very common among engineers to use "hard grease" on the crank pin and main journals, and it will very soon be used exclusively. With a good grade of grease your crank will not heat near so quickly as with oil and your engine will be much easier to keep clean; and if you are going to be an engineer be a neat one, keep your engine clean and keep yourself clean. You say you can't do that; but you can at least keep yourself respectable. You will most certainly keep your engine looking as though it had an engineer. Keep a good bunch of waste handy, and when it is necessary to wipe your hands use the waste and not your overalls, and when you go in to a nice dinner the cook will not say after you go out, "Look here where that dirty engineer sat." Now boys, these are things worth heeding. I have actually known threshing crews to lose good customers simply because of their dirty clothes. The women kicked and they had a right to kick. But to return to hard grease and suitable cups for same.

In attaching these grease cups on boxes not previously arranged for them, it would be well for you to know how to do it properly. You will remove the journal, take a gouge and cut a clean groove across the box, starting in at one corner, about 1/8 of an inch from the point of box and cut diagonally across coming out at the opposite corner on the other end of box. Then start at the opposite corner and run through as before, crossing the first groove in the center of box. Groove both halves of box the same, being careful not to cut out at either end, as this will allow the grease to escape from box and cause unnecessary waste. The chimming or packing in box should be cut so as to touch the journal at both ends of box, but not in the center or between these two points. So, when the top box is brought down tight, this will form another reservoir for the grease. If the box is not tapped directly in the center for cup, it will be necessary to cut other grooves from where it is tapped into the grooves already made. A box prepared in his way will require but little attention if you use good grease.

A HOT BOX

You will sometimes get a hot box. What is the best remedy? Well, I might name you a dozen, and if I did you would most likely never have one on hand when it was wanted. So will only give you one, and that is white lead and oil, and I want you to provide yourself with a can of this useful article. And should a journal or box get hot on your hands and refuse to cool with the usual methods, remove the cup, and after mixing a portion of the lead with oil, put a heavy coat of it on the journal, put back the cup and your journal will cool off very quickly. Be careful to keep all grit or dust out of your can of lead. Look after this part of it yourself. It is your business.

PART SEVEN

Before taking up the handling of a Traction Engine, we want to tell you of a number of things you are likely to do which you ought not to do.

Don't open the throttle too quickly, or you may throw the drive belt off, and are also more apt to raise the water and start priming.

Don't attempt to start the engine with the cylinder cocks closed, but make it a habit to open them when you stop; this will always insure your cylinder being free from water on starting.

Don't talk too much while on duty.

Don't pull the ashes out of ash pan unless you have a bucket of water handy.

Don't start the pump when you know you have low water.

Don't let it get low.

Don't let your engine get dirty.

Don't say you can't keep it clean.

Don't leave your engine at night till you have covered it up.

Don't let the exhaust nozzle lime up, and don't allow lime to collect where the water enters the boiler, or you may split a heater pipe or knock the top off of a check valve.

Don't leave your engine in cold weather without first draining all pipes.

Don't disconnect your engine with a leaky throttle.

Don't allow the steam to vary more than 10 or 15 pounds while at work.

Don't allow anyone to fool with your engine.

Don't try any foolish experiments on your engine.

Don't run an old boiler without first having it thoroughly tested.

Don't stop when descending a steep grade.

Don't pull through a stockyard without first closing the damper tight.

Don't pull onto a strange bridge without first examining it.

Don't run any risk on a bad bridge.

A TRACTION ENGINE ON THE ROAD

You may know all about an engine. You may be able to build one, and yet run a traction in the ditch the first jump.

It is a fact that some men never can become good operators of a traction engine, and I can't give you the reason why any more than you can tell why one man can handle a pair of horses better than another man who has had the same advantages. And yet if you do ditch your engine a few times, don't conclude that you can never handle a traction.

If you are going to run a traction engine I would advise you to use your best efforts to become an expert at it. For the expert will hook up to his load and get out of the neighborhood while the awkward fellow is getting his engine around ready to hook up.

The expert will line up to the separator the first time, while the other fellow will back and twist around for half an hour, and then not have a good job.

Now don't make the fatal mistake of thinking that the fellow is an expert who jumps up on his engine and jerks the throttle open and yanks it around backward and forward, reversing with a snap, and makes it stand-up on its hind wheels.

If you want to be an expert you must begin with the throttle, therein lies the secret of the real expert. He feels the power of his engine through the throttle. He opens it just enough to do what he wants it to do. He therefore has complete control of his engine. The fellow who backs his engine up to the separator with an open throttle and must reverse it to keep from running into and breaking something, is running his engine on his muscle and is entitled to small pay.

The expert brings his engine back under full control, and stops it exactly where he wants it. He handles his engine with his head and should be paid accordingly. He never makes a false move, loses no time, breaks nothing, makes no unnecessary noise, does not get the water all stirred up in the boiler, hooks up and moves out in the same quiet manner, and the onlookers think he could pull two such loads, and say he has a great engine, while the engineer of muscle would back up and jerk his engine around a half dozen times before he could make the coupling, then with a jerk and a snort he yanks the separator out of the holes, and the onlookers think he has about all he can pull.

Now these are facts, and they cannot be put too strong, and if you are going to depend on your muscle to run your engine, don't ask any more money than you would get at any other day labor.

You are not expected to become an expert all at once. Three things are essential to be able to handle a traction engine as it should be handled.

First, a thorough knowledge of the throttle. I don't mean that you should simply know how to pull it open and shut it. Any boy can do that. But I mean that you should be a good judge of the amount of power it will require to do what you may wish to do, and then give it the amount of throttle that it will require and no more. To illustrate this I will give an instance.

An expert was called a long distance to see an engine that the operator said would not pull its load over the hills he had to travel.

The first pull he had to make after the expert arrived was up the worst hill he had. When he approached the grade he threw off the governor belt, opened the throttle as wide as he could get it, and made a run for the hill. The result was, that he lifted the water and choked the engine down before he was half way up. He stepped off with the remark, "That is the way the thing does." The expert then locked the hind wheels of the separator with a timber, and without raising the pressure a pound, pulled it over the hill. He gave it just throttle enough to pull the load, and made no effort to hurry ii, and still had power to spare.

A locomotive engineer makes a run for a hill in order that the momentum of his train will help carry him over. It is not so with a traction and its load; the momentum that you get don't push very hard.

The engineer who don't know how to throttle his engine never knows what it will do, and therefore has but little confidence in it; while the engineer who has a thorough knowledge of the throttle and uses it, always has power to spare and has perfect confidence in his engine. He knows exactly what he can do and what he cannot do.

The second thing for you to know is to get onto the tricks of the steer wheel.

This will come to you naturally, and it is not necessary for me to spend much time on it. All new beginners make the mistakes of turning the wheel too often. Remember this-that every extra turn to the right requires two turns to the left, and every extra turn to the left requires two more to the right; especially is this the care if your engine is fast on the road.

The third thing for you to learn, is to keep your eyes on the front wheels of your engine, and not be looking back to see if your load in coming.

In making a difficult turn you will find it very much to your advantage to go slow, as it gives you much better control of your front wheels, and it is not a bad plan for a beginner to continue to go slow till he has perfect confidence in his ability to handle the steer wheel as it may keep you out of some bad scrapes.

How about getting into a hole? Well, you are not interested half as much in knowing how to get into a hole as You are in knowing how to get out. An engineer never shows the stuff he is made of to such good advantage as when he gets into a hole; and he is sure to get there, for one of the traits of a traction engine is its natural ability to find a soft place in the ground.

Head work will get you out of a bad place quicker than all the steam you can get in your boiler. Never allow the drivers to turn without doing some good. If you are in a hole, and you are able to turn your wheels, you are not stuck; but don't allow your wheels to slip, it only lets you in deeper. If your wheels can't get a footing, you want to give them something to hold to. Most smart engineers will tell you that the best thing is a heavy chain. That is true. So are gold dollars the best things to buy bread with, but you have not always got the gold dollars, neither have you always got the chain. Old hay or straw is a good thing; old rails or timber of any kind. The engineer with a head spends more time trying to give his wheels a hold than he does trying to pull out, while the one without a head spends more time trying to pull out than he does trying to secure a footing, and the result is, that the first fellow generally gets out the first attempt, while the other fellow is lucky if he gets out the first half day.

If you have one wheel perfectly secure, don't spoil it by starting your engine till you have the other just as secure.

If you get into a place where your engine is unable to turn its wheels, then your are stuck, and the only thing for you to do is to lighten your load or dig out. But under all circumstances your engine should be given the benefit of your judgment.

All traction engines to be practical must of a necessity, be reversible. To accomplish this, the link with the double eccentric is the one most generally used, although various other devices are used with more or less success. As they all accomplish the same purpose it is not necessary for us to discuss the merits or demerits of either.

The main object is to enable the operator to run his engine either backward or forward at will, but the link is also a great cause of economy, as it enables the engineer to use the steam more or less expansively, as he may use more or less power, and, especially is this true, while the engine is on the road, as the power required may vary in going a short distance, anywhere from nothing in going down hill, to the full power of your engine in going up.

By using steam expansively, we mean the cutting off of the steam from the cylinder, when the piston has traveled a certain part of its stroke. The earlier in the stroke this is accomplished the more benefit you get of the expansive force of the steam.

The reverse on traction engines is usually arranged to cut off at I/4, I/2 or 3/4. To illustrate what is meant by "cutting off" at I/4, I/2 or 3/4, we will suppose the engine has a I2 inch stroke. The piston begins its stroke at the end of cylinder, and is driven by live steam through an open port, 3 inches or one quarter of the stroke, when the port is closed by the valve shutting the steam from the cylinder, and the piston is driven the remaining 9 inches of its stroke by the expansive force of the steam. By cutting off at I/2 we mean that the piston is driven half its stroke or 6 inches by live steam, and by the expansion of the steam the remaining 6 inches; by 3/4 we mean that live steam is used 9 inches before cutting off, and expansively the remaining 3 inches of stroke.

Here is something for you to remember: "The earlier in the stroke you cut off the greater the economy, but less the power; the later you cut off the less the economy and greater the power."

Suppose we go into this a little farther. If you are carrying I00 pounds pressure and cut off at I/4, you can readily see the economy of fuel and water, for the steam is only allowed to enter the cylinder during I/4 of its stroke; but by reason of this, you only get an average pressure on the piston head of 59 pounds throughout the stroke.

But if this is sufficient to do the work, why not take advantage of it and thereby save your fuel and water? Now, with the same pressure as before, and cutting off at I/2, you have an average pressure on piston head of 84 pounds, a loss of 50 per cent in economy and a gain of 42 per cent in power. Cutting of at 3/4 gives you an average pressure of 96 pounds throughout the stroke. A loss on cutting off at I/4 of 75 per cent in economy, and a gain of nearly 63 per cent in power. This shows that the most available point at which to work steam expansively is at I/4, as the percentage of increase of power does not equal the percentage of loss in economy. The nearer you bring the reverse lever to center of quadrant, the earlier will the valve cut the steam and the less will be the average pressure, while the farther away from the center the later in the stroke will the valve cut the steam, and the greater the average pressure, and, consequently, the greater the power. We have seen engineers drop the reverse back in the last notch in order to make a hard pull, and were unable to tell why they did so.

Now, as far as doing the work is concerned, it is not absolutely necessary that you know this; but if you do know it, you are more likely to profit by it and thereby get the best results out of your engine. And as this is our object, we want you to know it, and be benefitted by the knowledge. Suppose you are on the road with your engine and load, and you have a stretch of nice road. You are carrying a good head of steam and running with lever back in the corner or lower notch. Now your engine will travel along its regular speed, and say you run a mile this way and fire twice in making it. You now ought to be able to turn around and go back on the same road with one fire by simply hooking the lever up as short as it will allow to do the work. Your engine will make the same time with half the fuel and water, simply because you utilize the expansive force of the steam instead of using the live steam from boiler. A great many good engines are condemned and said to use too much fuel, and all because the engineer takes no pains to utilize the steam to the best advantage.

I have already advised you to carry a "high pressure;" by a high pressure I mean any where from I00 to I25 lbs. I have done this expecting you to use the steam expansively whenever possible, and the expansive force of steam increases very rapidly after you have reached 70 lbs. Steam at 80 lbs. used expansively will do nine times the work of steam at 25 lbs. Note the difference. Pressure 3 I-5 times greater.

Work performed, 9 times greater. I give you these facts trusting that you will take advantage of them, and if your engine at 100 or 100 lbs. will do your work cutting off at 1/4, don't allow it to cut off at 1/2. If cutting off at 1/2 will do the work, don't allow it to cut off at 3/4, and the result will be that you will do the work with the least possible amount of fuel, and no one will have any reason to find fault with you or your engine.

Now we have given you the three points which are absolutely necessary to the successful handling of a traction engine, We went through it with you when running as a stationary; then we gave you the pointers-to be observed when running as a traction or road engine. We have also given you hints on economy, and if you do not already know too much to follow our advice, you can go into the field with an engine and have no fears as to the results.

How about bad bridges?

Well, a bad bridge is a bad thing, and you cannot be too careful. When you have questionable bridges to cross over, you should provide yourself with good hard-wood planks. If you can have them sawed to order have them 3 inches in the center and tapering to 2 inches at the ends. You should have two of these about 16 feet long, and two 2x12 planks about 8 feet long. The short ones for culverts, and for helping with the longer ones in crossing longer bridges.

An engine should never be allowed to drop from a set of planks down onto the floor of bridge. This is why I advocate four planks. Don't hesitate to use the plank. You had better plank a dozen bridges that don't need it than to attempt to cross one that does need it. You will also find it very convenient to carry at least 50 feet of good heavy rope. Don't attempt to pull across a doubtful bridge with the separator or tank hooked directly to the engine. It is dangerous. Here is where you want the rope. An engine should be run across a bad bridge very slowly and carefully, and not allowed to jerk. In extreme cases it is better to run across by hand; don't do this but once; get after the road supervisors.

SAND.

An engineer wants a sufficient amount of "sand," but he don't want it in the road. However, you will find it there and it is the meanest road you will have to travel. A bad sand road requires considerable sleight of hand on the part of the engineer if he wishes to pull much of a load through it. You will find it to your advantage to keep your engine as straight as possible, as you are not so liable to start one wheel to slipping any sooner than the other. Never attempt to "wiggle" through a sand bar, and don't try to hurry through; be satisfied with going slow, just so you are going. An engine will stand a certain speed through sand, and the moment you attempt to increase that speed, you break its footing, and then you are gone. In a case of this kind, a few bundles of hay is about the best thing you can use under your drivers in order to get started again. But don't loose your temper; it won't help the sand any.

Now no doubt the reader wonders why I have said nothing about compound engines. Well in the first place, it is not necessary to assist you in your work, and if you can handle the single cylinder engine, you can handle the compound.

The question as to the advantage of a compound engine is, or would be an interesting one if we cared to discuss it.

The compound traction engine has come into use within the past few years, and I am inclined to think more for sort of a novelty or talking point rather than to produce a better engine. There is no question but that there is a great advantage in the compound engine, for stationary and marine engines.

In a compound engine the steam first enters the small or high pressure cylinder and is then exhausted into the large or low pressure cylinder, where the expansive force is all obtained.

Two cylinders are used because we can get better results from high pressure in the use of two cylinders of different areas than by using but one cylinder, or simple engine.

That there is a gain in a high pressure, can be shown very easily:

For instance, 100 pounds of coal will raise a certain amount of water from 60 degrees, to 5 pounds steam pressure, and 102.9 pounds would raise the same water to 80 pounds, and 104.4 would raise it to 160 pounds, and this 160 pounds would produce a large increase of power over the 80 pounds at a very slight increase of fuel. The compound engine will furnish the same number of horse power, with less fuel than the simple engine, but only when they are run at the full load all the time.

If, however, the load fluctuates and should the load be light for any considerable part of the day, they will waste the fuel instead of saving it over the simple engine.

No engine can be subjected to more variation of loads than the traction engine, and as the above are facts the reader can draw his own conclusions.

FRICTION CLUTCH

The friction clutch is now used almost exclusively for engaging the engine with the propelling gearing of the traction drivers, and it will most likely give you more trouble than any one thing on your engine, from the fact that to be satisfactory they require a nicety of adjustment, that is very difficult to attain, a half turn of the expansion bolt one way or the other may make your clutch work very nicely, or very unsatisfactory, and you can only learn this by carefully adjusting of friction shoes, until you learn just how much clearance they will stand when lever is out, in order to hold sufficient when lever is thrown in. If your clutch fails to hold, or sticks, it is not the fault of the clutch, it is not adjusted properly. And you may have it correct today and tomorrow it will need readjustment, caused by the wear in the shoes; you will have to learn the clutch by patience and experience.

But I want to say to you that the friction clutch is a source of abuse to many a good engineer, because the engineer uses no judgment in its use.

A certain writer on engineering makes use of the following, and gives me credit: "Sometimes you may come to an obstacle in the road, over which your engine refuses to go, you may perhaps get over it in this way, throw the clutch-lever so as to disconnect the road wheels, let the engine get up to full speed and then throw the clutch level back so as to connect the road wheels." Now I don't thank any one for giving me credit for saying any such thing. That kind of thing is the hight of abuse of an engine.

I am aware that when the friction clutch first came into use, their representatives made a great talk on that sort of thing to the green buyer. But the good engineer knows better than to treat his engine that way.

Never attempt to pull your loads over a steep hill without being certain that your clutch is in good shape, and if you have any doubts about it put in the tight gear pin. Most all engines have both the friction and the tight gear pin. The pin is much the safer in a hilly country, and if you have learned the secret of the throttle you can handle just as big load with the pin as with the clutch, and will never tear your gearing off or lose the stud bolts in boiler.

The following may assist you in determining or arriving at some idea of the amount of power you are supplying with your engine:

For instance, a I inch belt of the standard grade with the proper tention, neither too tight or too loose, running at a. maximum spead of 800 ft. a minute will transmit one horse power, running 1600 ft. 2 horse power and 2400 ft. 3 horse power. A 2 inch belt, at the same speed, twice the power.

Now if you know the circumference of your fly wheel, the number of revolutions your engine is making and the width of belt, you can figure very nearly the amount of power you can supply without slipping your belt. For instance, we will say your fly wheel is 40 inches in diameter or 10.5 feet nearly in circumference and your engine was running 225 revolutions a minute, your belt would be traveling 225 x 10.5 feet = 2362.5 feet or very nearly 2400 ft. and if I inch of belt would transmit 3 H. P. running this speed, a 6 inch belt would transmit 18 H.P., a 7 inch belt, 21 H.P., an 8 inch belt 24 H.P., and so on. With the above as a basis for figuring you can satisfy yourself as to the power you are furnishing. To get the best results a belt wants to sag slightly as it hugs the pulley closer, and will last much longer.

SOMETHING ABOUT SIGHT-FEED LUBRICATORS

All such lubricators feed oil through the drop-nipple by hydrostatic pressure; that is, the water of condensation in the condenser and its pipe being elevated above the oil magazine forces the oil out of the latter by just so much pressure as the column of water is higher than the exit or outlet of oil-nipple. The higher the column of water the more positive will the oil feeds. As soon as the oil drop leaves the nipple it ceases to be actuated by the hydrostatic pressure, and rises through the water in the sight-glass merely by the difference of its specific gravity, as compared with water and then passes off through the ducts provided to the parts to be lubricated.

For stationary engines the double connection is preferable, and should always be connected to the live steam pipe above the throttle. The discharge arm should always be long enough (4 to 6 inches) to insure the oil magazine and condenser from getting too hot, otherwise it will not condense fast enough to give continuous feed of oil. For traction or road engines the single connection is used. These can be connected to live steam pipe or directly to steam chest.

In a general way it may be stated that certain precaution must be taken to insure the satisfactory operation of all sight-feed lubricators. Use only the best of oil, one gallon of which is worth five gallons of cheap stuff and do far better service, as inferior grades not only clog the lubricator but chokes the ducts and blurs the sight-glass, etc., and the refuse of such oil will accumulate in the cylinder sufficiently to cause damage and loss of power, far exceeding the difference in cost of good oil over the cheap grades.

After attaching a lubricator, all valves should be opened wide and live steam blown through the outer vents for a few minutes to insure the openings clean and free. Then follow the usual directions given with all lubricators. Be particular in getting your lubricator attached so it will stand perfectly plum, in order that the drop can pass up through the glass without touching the sides, and keep the drop-nipple clean, be particular to drain in cold weather.

Now, I am about to leave you alone with your engine, just as I have left any number of young engineers after spending a day with them in the field and on the road. And I never left one, that I had not already made up my mind fully, as to what kind of an engineer he would make.

TWO WAYS OF READING

Now there are two ways to read this book, and if I know just how you had read it I could tell you in a minute whether to take hold of an engine or leave it alone. If you have read it one way, you are most likely to say "it is no trick to run an engine." If you have read it the other way you will say, "It is no trouble to learn how to run an engine." Now this fellow will make an engineer, and will be a good one. He has read it carefully, noting the drift of my advice. Has discovered that the engineer is not expected to build an engine, or to improve it after it has been built. Has recognized the fact that the principle thing is to attend to his own business and let other people attend to theirs. That a monkey wrench is a tool to be left in the tool box till he knows he needs it. That muscle is a good thing to have but not necessary to the successful engineer. That an engineer with a bunch of waste in his hand is a better recommendation than an "engineer license." That good common sense, and a cool head is the very best tools he can have. Has learned that carelessness will get him into trouble, and that to "forget" costs money.

Now the fellow who said "It is no trick to run an engine," read this book another way. He did not see the little points. He was hunting for big theories, scientific theories, something he could not understand, and didn't find them. He expected to find some bright scheme to prevent a boiler from exploding, didn't notice the simple little statement, "keep water in it," that was too commonplace to notice. He was looking for cuts, diagrams, geometrical figures, theories for constructing engines and boilers and all that sort of thing and didn't find them. Hence "It is no trick to run an engine."

If this has been your idea of "Rough and Tumble Engineering" forget all about your theory, and go back and read it over and remember the little suggestions and don't expect this book to teach you how to build an engine. We didn't start out to teach you anything of the kind. That is a business of itself. A good engineer gets better money than the man who builds them. Read it as if you wanted to know how to run an engine and not how to build one.

Study the following questions and answers carefully. Don't learn them like you would a piece of poetry, but study them, see if they are practical; make yourself thoroughly acquainted with the rule for measuring the horse-power of an engine; make yourself so familiar with it that you could figure any engine without referring to the book. Don't stop at this, learn to figure the heating surface in any boiler. It will enable you to satisfy yourself whether you are working your boiler or engine too hard or what it ought to be capable of doing.

SOME THINGS TO KNOW

Q. What is fire? A. Fire is the rapid combustion or consuming of organic matter.

Q. What is water? A. Water is a compound of oxygen and hydrogen. In weight 88 9-10 parts oxygen to II I-I0 hydrogen. It has its maximum density at 39 degrees Fahr., changes to steam at 2I2 degrees, and to ice at 32 degrees.

Q. What is smoke? A. It is unconsumed carbon finely divided escaping into open air.

Q. Is excessive smoke a waste of fuel? A. Yes.

Q. How will you prevent it A. Keep a thin fire, and admit cold air sufficient to insure perfect combustion.

Q. What is low water as applied to a boiler? A. It is when the water is insufficient to cover all parts exposed to the flames.

Q. What is the first thing to do on discovering that you have low water? A. Pull out the fire.

Q. Would it be safe to open the safety valve at such time? A. No.

Q. Why not? A. It would relieve the pressure on the water which being allowed to flow over the excessive hot iron would flash into steam, and might cause an explosion.

Q. Why do boilers sometimes explode just on the point of starting the engine? A. Because starting the engine has the same effect as opening the safety valve.

Q. Are there any circumstances under which an engineer is justified in allowing the water to get low? A. No.

Q. Why do they sometimes do it? A. From carelessness or ignorance.

Q. May not an engineer be deceived in the gauge of water? A. Yes.

Q. Is he to be blamed under such circumstances? A. Yes.

Q. Why? A. Because if he is deceived by it it shows he has neglected something.

Q. What is meant by "Priming." A. It is the passing of water in visible quantities into the cylinder with the steam.

Q. What would you consider the first duty of an engineer on discovering that the water was foaming or priming A. Open the cylinder cocks at once, and throttle the steam.

Q. Why would you do this? A. Open the cocks to enable the water to escape, and throttle the steam so that the water would settle.

Q. Is foaming the same as priming? A. Yes and no.

Q. How do you make that out? A. A boiler may foam without priming, but it can't prime without first foaming..

Q. Where will you first discover that the water is foaming? A. It will appear in the glass gauge, the glass will have a milky appearance and the water will seem to be running down from the top, There will be a snapping or cracking in the cylinder as quick as priming begins.

Q. What causes a boiler to foam? A. There are a number of causes. It may come from faulty construction of boiler; it may have insufficient steam room. It may be, and usually is, from the use of bad water, muddy or stagnant water, or water containing any soapy substance.

Q. What would you do after being bothered in this way? A. Clean out the boiler and get better water if possible.

Q. How would you manage your pumps while the water was foaming. A. Keep them running full.

Q. Why? A. In order to make up for the extra amount of water going out with the steam.

Q. What is "cushion?" A. Cushion is steam retained or admitted in front of the piston head at the finish of stroke, or when the engine is on "center."

Q. What is it for? A. It helps to overcome the "inertia" and momentum of the reciprocating parts of the engine, and enables the engine to pass the center without a jar.

Q. How would you increase the cushion in an engine? A. By increasing the lead.

Q. What is lead? A. It is the amount of opening the port shows on steam end of cylinder when the engine is on dead center.

Q. Is there any rule for giving an engine the proper lead? A. No.

Q. Why not? A. Owing to their variation in construction, speed, etc.

Q. What would you consider the proper amount of lead, generally. A. From I/32 to I/I6.

Q. What is "lap?" A. It is the distance the valve overlaps the steam ports when in mid position.

Q. What is lap for? A. In order that the steam may be worked expansively.

Q. When does expansion occur in a cylinder? A. During the time between which the port closes and the point at which the exhaust opens.

Q. What would be the effect on an engine if the exhaust opened too soon? A. It would greatly lessen the power of the engine.

Q. What effect would too much lead have. A. It would also weaken the engine, as the steam would enter before the piston had reached the end of the stroke, and would tend to prevent it passing the center.

Q. What is the stroke of an engine? A. It is the distance the piston travels in the cylinder.

Q. How do you find the speed of a piston per minute? A. Double the stroke and multiply it by the number of revolutions a minuet. Thus an engine with a 12 inch stroke would travel 24 inches, or 2 feet, at a revolution. If it made 200 revolutions a minute, the travel of piston would be 400 feet a minute.

Q. What is considered a horse power as applied to an engine? A. It is power sufficient to lift 33,000 pounds one foot high in one minute.

Q. What is the indicated horse power of an engine? A. It is the actual work done by the steam in the cylinder as shown by an indicator.

Q. What is the actual horse power? A. It is the power actually given off by the driving belt and pulley.

Q. How would you find the horse power of an engine? A. Multiply the area of the piston by the average pressure, less 5; multiply this product by the number of feet the piston travels per minute; divide the product by 33,000; the result will be horse power of the engine.

Q. How will you find the area of piston? A. Square the diameter of piston and multiply it by .7854.

Q. What do you mean by squaring the diameter? A. Multiplying it by itself. If a cylinder is 6 inches in diameter, 36 multiplied by .7854, gives the area in square inches.

Q. What do you mean by average pressure? A. If the pressure on boiler is 60 pounds, and the engine is cutting off at 1/2 stroke, the pressure for the full stroke would be 50 pounds.

Q. Why do you say less 5 pounds? A. To allow for friction and condensation.

Q. What is the power of a 7 x 10 engine, running 200 revolutions, cutting off at 1/2 stroke with 60 pounds steam? A. 7 x 7 = 49 x .7854 = 38.4846. The average pressure of 60 pounds would be 50 pounds less 5 = 45 pounds; 38-4846 x 45 = 1731.8070 x .333 1/3, (the number of feet the piston travels per minute) 577,269.0000 by 33,000=17 1/2 horse power.

Q. What is a high pressure engine? A. It is an engine using steam at a high pressure and exhausting into the open air.

Q. What is a low pressure engine? A. It is one using steam at a low pressure and exhausting into a condenser, producing a vacuum, the piston being under steam pressure on one side and vacuum on the other.

Q. What class of engines are farm engines? A. They are high pressure.

Q. Why? A. They are less complicated and less expensive.

Q. What is the most economical pressure to carry on high pressure engine? A. From 90 to 110 pounds.

Q. Why is high pressure more economical than low pressure? A. Because the loss is greater in low pressure owing to the atmospheric pressure. With 45 pounds steam the pressure from the atmosphere is 15 pounds, or 1/3, leaving only 30 pounds of effective power; while with 90 pounds the atmospheric pressure is only 1-6 of the boiler pressure.

Q. Does it require any more fuel to carry I00 pounds than it does to carry 60 pounds? A. It don't require quite as much.

Q. If that is the case why not increase the pressure beyond this and save more fuel? A. Because we would soon pass the point of safety in a boiler, and the result would be the loss of life and property.

Q. What do you consider a safe working pressure on a boiler? A. That depends entirely on its diameter. While a boiler of 30 inches in diameter 3/8 inch iron would carry 140 pounds, a boiler of the same thickness 80 inches in diameter would have a safe working pressure of only 50 pounds, which shows that the safe working pressure decreases very rapidly as we increase the diameter of boiler. This is the safe working pressure for single riveted boilers of this diameter. To find the safe working pressure of a double riveted boiler of same diameter multiply the safe pressure of the single riveted by 70, and divide by 56, will give a safe pressure of a double riveted boiler.

Q. Why is a steel boiler superior to an iron boiler? A. Because it is much lighter and stronger.

Q. Does boiler plate become stronger or weaker as it becomes heated? A. It becomes tougher or stronger as it is heated, till it reaches a temperature Of 550 degrees when it rapidly decreases its power of resistance as it is heated beyond this temperature.

Q. How do you account for this? A. Because after you pass the maximum temperature of 550 degrees, the more you raise the temperature the nearer you approach its fusing point when its tenacity or resisting power is nothing.

Q. What is the degree of heat necessary to fuse iron? A. 2912 degrees.

Q. Steel? A. 2532 degrees.

Q. What class of boilers are generally used in a threshing engine? A. The flue boiler and the tubular boiler.

Q. About what amount of heating and grate surface is required per horse power in a flue boiler. A. About 15 square feet of heating surface and 3/4 square feet of grate surface.

Q. What would you consider a fair evaporation in a flue boiler? A. Six pounds of water to 1 pound of coal.

Q. How do these dimensions compare in a tubular boiler. A. A tubular boiler will require 1/4 less grate surface, and will evaporate about 8 pounds of water to 1 pound of coal.

Q. Which do you consider the most available? A. The tubular boiler.

Q. Why? A. It is more economical and is less liable to "collapse?"

Q. What do you mean by "collapse?" A. It is a crushing in of a flue by external pressure.

Q. Is a tube of a large diameter more liable to collapse than one of small diameter? A. Yes.

Q. Why? A. Because its power of resistance is much less than a tube of small diameter.

Q. Is the pressure on the shell of a boiler the same as on the tubes? A. No.

Q. What is the difference? A. The shell of boiler has a tearing or internal pressure while the tubes have a crushing or external pressure.

Q. What causes an explosion? A. An explosion occurs generally from low water, allowing the iron to become overheated and thereby weakened and unable to withstand the pressure.

Q. What is a "burst?" A. It is that which occurs when through any defect the water and steam are allowed to escape freely without further injury to boiler.

Q. What is the best way to prevent an explosion or burst? A. (I) Never go beyond a safe working pressure. (2) Keep the boiler clean and in good repair. (3) Keep the safety valves in good shape and the water at its proper height.

Q. What is the first thing to do on going to your engine in the morning? A. See that the water is at its proper level.

Q. What is the proper level? A. Up to the second gauge.

Q. When should you test or try the pop valve? A. As soon as there is a sufficient pressure.

Q. How would you start your engine after it had been standing over night? A. Slowly.

Q. Why? A. In order to allow the cylinder to become hot, and that the water or condensed steam may escape without injury to the cylinder.

Q. What is the last thing to do at night? A. See that there is plenty of water in boiler, and if the weather is cold drain all pipes.

Q. What care should be taken of the fusable plug? A. Keep it scraped clean, and not allow it to become corroded on top.

Q. What is a fusible plug? A. It is a hollow cast plug screwed into the crown sheet or top of fire box, and having the hollow or center filled with lead or babbit.

Q. Is such a plug a protection to a boiler? A. It is if kept in proper condition.

Q. Can you explain the principle of the fusible or soft plug as it is sometimes called? A. It is placed directly over the fire, and should the water fall below the crown sheet the lead fuses or melts and allows the steam to flow down on top of the fire, destroys the heat and prevents the burning of crown sheet.

Q. Why don't the lead fuse with water over it? A. Because the water absorbs the heat and prevents it reaching the fusing point.

Q. What is the fusing point of lead? A. 618 degrees.

Q. Is there any objection to the soft plug? A. There is, in the hands of some engineers.

Q. Why? A. It relieves him of the fear of a dry crown sheet, and gives him an apparent excuse for low water.

Q. Is this a real or legitimate objection? A. It is not.

Q. What are the two distinct classes of boilers? A. The externally and internally fired boilers.

Q. Which is the most economical? A. The internally fired boiler.

Q. Why? A. Because the fuel is all consumed in close contact with the sides of furnace and the loss from radiation is less than in the externally fired.

Q. To what class does the farm or traction engine belong? A. To the internally fired.

Q. How would you find the H.P. of such a boiler? A. Multiply in inches the circumference or square of furnace, by its length, then multiply, the circumference of one tube by its total length, and this product by the number of tubes also taking into account the surface in tube sheet, add these products together and divide by 144, this will give you the number of square feet of heating surface in boiler. Divide this by 14 or 15 which will give the H.P. of boiler.

Q. Why do you say 14 or 15? A. Because some claim that it requires 14 feet of heating surface to the H.P. and others 15. To give you my personal opinion I believe that any of the standard engines today with good coal and properly handled, will and are producing 1 H.P. for as low as every 10 feet of surface. But to be on the safe side it is well to divide by 15 to get the H.P. of your boiler, when good and bad fuel is considered.

Q. How would you find the approximate weight of a boiler by measurement? A. Find the number of square feet in surface of boiler and fire box, and as a sheet of boiler iron or steel 1/16 of an inch thick, and one foot square, weighs 2.52 pounds, would multiply the number of square feet by 2.52 and this product by the number of 16ths or thickness of boiler sheet, which would give the approximate, or very near the weight of the boiler.

Q. What would you recognize as points in a good engineer. A. A good engineer keeps his engine clean, washes the boiler whenever he thinks it needs it. Never meddles with his engine, and allows no one else to do so. Goes about his work quietly, and is always in his place, only talks when necessary, never hammers or bruises any part of his engine, allows no packing to become baked or burnt in the stuffing box or glands, renews them as quick as they show that they require it. Never neglects to oil, and then uses no more than is necessary. He carries a good gauge of water and a uniform pressure of steam. He allows no unusual noise about his engine to escape his notice he has taught his ear to be his guide. When a job is about finished you will see him cleaning his ash pan, getting his tools together, a good fire in fire box, in fact all ready to go, and he looses no time after the belt is thrown off. He hooks up to his load quietly, and is the first man ready to go.

*Q. When the piston head is in the exact center of cylinder, is the engine on the quarter? *A. It is supposed to be, but is not.

*Q. Why not? A. The angularity of the rod prevents it reaching the quarter.

*Q. Then when the engine is on the exact quarter what position does the piston head occupy? A. It is nearest the end next to crank.

Q. If this is the case, which end of cylinder is supposed to be the stronger? A. The opposite end, or end furtherest from crank.

Q. Why? A. Because this end gets the benefit of the most travel, and as it makes it in the same time, it must travel faster.

*Q. At what part of the cylinder does the piston head reach the greatest speed? A. At and near the center.

*Q. Why? Figure this out for yourself. *Note. The above few questions are given for the purpose of getting you to notice the little peculiarities of the crank engine, and are not to be taken into consideration in the operation of the same.

Q. If you were on the road and should discover that you had low water, what would you do? A. I would drop my load and hunt a high place for the front end of my engine, and would do it quickly to.

Q. If by some accident the front end of your engine should drop down allowing the water to expose the crown sheet, what would you do? A. If I had a heavy and hot fire, would shovel dirt into the fire and smother it out.

Q. Why would you prefer this to drawing the fire? A. Because it would reduce the heat at once, instead of increasing it for a few minutes while drawing out the hot bed of coals, which is a very unpleasant job.

Q. Would you ever throw water in the fire box? A. No. It might crack the side sheets, and would most certainly start the flues.

Q. You say, in finding low water while on the road, you would run your engine with the front end on high ground. Why would you do this? A. In order that the water would raise over the crown sheet, and thus make it safe to pump up the water.

Q. While your engine was in this shape would you not expose the front end of flues'? A. Yes, but as the engine would not be working this would do no damage.

Q. If you were running in a hilly country how would you manage the boiler as regards water? A. Would carry as high as the engine would allow, without priming.

Q. Suppose you had a heavy load or about all you could handle, and should approach a long steep hill, what condition should the water and fire be to give you the most advantage? A. A moderately low gauge of water and a very hot fire.

Q. Why a moderately low gauge of water? A. Because the engine would not be so liable to draw the water or prime in making the hard pull.

Q. Why a very hot fire? A. So I could start the pumps full without impairing or cutting the pressure.

Q. When would you start your pump? A. As soon as fairly started up the hill.

Q. Why? A. As most hills have two sides, I would start them full in order to have a safe gauge to go down, without stoping to pump up.

Q. What would a careful engineer do before starting to pull a load over a steep hill? A. He would examine his clutch, or gear pin.

Q. How would you proceed to figure the road speed of traction. A. Would first determine the circumference of driver, then ascertain how many revolutions the engine made to one of the drivers. Multiply the number of revolutions the engine makes per minute by 60, this will give the number of revolutions of engine per hour. Divide this by the number of revolutions the engine makes to the drivers once, and this will give you the number of revolutions the drivers will make in one hour, and multiplying this by the circumference of driver in feet, and it will tell you how many feet your engine is traveling per hour, and this divided by 5280, the number of feet in a mile, would tell you just what speed your engine would make on the road.

THINGS HANDY FOR THE ENGINEER

The first edition of this work brought me a great many letters asking where certain articles could be procured, what I would recommend, etc. These questions required attention and as the writers had bought and paid for their book it was due them that they get the benefit of my experience, as nothing is so discouraging to the young engineer as to be continually annoyed by unreliable and inferior fittings used more or less on all engines. I have gone over my letter file and every article asked for will be taken up in the order, showing the relative importance of each article in the minds of engineers. For instance, more letters reached me asking for a good brand of oil than any other one article. Then comes injectors, lubricators have third place, and so on down the list. Now without any intention of advertising anybody's goods I will give you the benefit of my years of experience and will be very careful not to mention or recommend anything which is not strictly first class, at least so in my opinion, and as good as can be had in its class, yet in saying that these articles are good does not say that others are not equally as good. I am simply anticipating the numerous letters I otherwise would receive and am answering them in a lump bunch. If you have no occasion to procure any of these articles, the naming of them will do no harm, but should you want one or more you will make no mistake in any one of them.

OIL

As I have stated, more engineers asked for a good brand of oil than for any other one article and I will answer this with less satisfaction to myself than any other for this reason: You may know what you want, but you do not always get what you call for. Oil is one of those things that cannot be branded, the barrel can, but then it can be filled with the cheapest stuff on the market. If you can get Capital Cylinder Oil your valve will give you no trouble. If you call for this particular brand and it does not give you satisfaction don't blame me or the oil, go after the dealer; he did not give you what you called for. The same can be said of Renown Engine Oil. If you can always have this oil you will have no fault to find with its wearing qualities, and it will not gum on your engine, but as I have said, you may call for it and get something else. If your valve or cylinder is giving you any trouble and you have not perfect confidence in the dealer from whom you usually get your cylinder oil send direct to The Standard Oil Company for some Capital Cylinder Oil and you will get an oil that will go through your cylinder and come out the exhaust and still have some staying qualities to it. The trouble with so much of the so called cylinder oil is that it is so light that the moment it strikes the extreme heat in the steam chest it vaporizes and goes through the cylinder in the form of vapor and the valve and cylinder are getting no oil, although you are going through all the necessary means to oil them.

It is somewhat difficult to get a young engineer to understand why the cylinder requires one grade of oil and the engine another. This is only necessary as a matter of economy, cylinder or valve oil will do very well on the engine, but engine oil will not do for the cylinder. And as a less expensive oil will do for the engine we therefore use two grades of oil.

Engine oil however should be but little lower in quality than the cylinder oil, owing to the proximity of the bearings to the boiler, they are at all times more or less heated, and require a much heavier oil than a journal subject only to the heat of its own friction. The Renown Engine Oil has the peculiarity of body or lasting qualities combined with the fact that it does not gum on the hot iron and allows the engine to be wiped clean.

INJECTORS

The next in the list of inquiries was for a reliable injector. I was not surprised at this for up to a few years ago there were a great many engines running throughout the country with only the independent or cross-head pump, and engineers wishing to adopt the injector naturally want the best, while others had injectors more or less unsatisfactory. In replying to these letters I recommend one of three or four different makes (all of which I had found satisfactory) with a request that the party asking for same should write to me if the injector proved unsatisfactory in any way. Of all the letters received, I never got one stating any objection to either the Penberthy or the Metropolitan. This fact has led me to think that probably my reputation as a judge of a good article was safer by sticking to the two named, which I shall do until I know there is something better. This does not mean that there are not other good injectors, but I am telling you what I know to be good, and not what may be good. The fact that I never received a single complaint from either of them was evidence to me that the makers of these two injectors are very careful not to allow any slighting of the work. They therefore get out no defective injectors. The Penberthy is made by The Penberthy Injector Co., of Detroit, Mich., and the Metropolitan by The Hayden & Derby Mfg. Co., New York, N. Y.

SIGHT FEED LUBRICATOR

These come next in the long list of inquiries and wishing to satisfy myself as to the relative superiority of various cylinder Lubricators, I resorted to the same method as persued in regard to injectors. This method is very satisfactory to me from the fact that it gives us the actual experience of a class of engineers who have all conditions with which to contend, and especially the unfavorable conditions. I have possibly written more letters in answer to such questions as: "Why my Lubricator does this or that; and why it don't do so and so?" than of any other one part of an engine, (as a Sight Feed Lubricator might in this day be considered a part of an engine.) Of all the queries and objections made of the many Lubricators, there are two showing the least trouble to the operator. There are the Wm. Powell Sight Feed Lubricator (class "A") especially adapted to traction and road engines owing to the sight-glass being of large diameter, which prevents the drop touching the side of glass, while the engine is making steep grades and rough uneven roads, made by The Wm. Powell Co., Cincinnati, O., and for sale by any good jobbing house, and the Detroit Lubricator made by the Detroit Lubricator Co., of Detroit, Mich. I have never received a legitimate objection to either of these two Lubricators, but I received the same query concerning both, and this objection, if it may be called such, is so clearly no fault of the construction or principle of the Lubricator that I have concluded that they are among if not actually the best sight feed Lubricator on the market to-day. The query referred to was: "Why does my glass fill with oil?" Now the answer to this is so simple and so clearly no fault of the Lubricator that I am entirely satisfied that by recommending either of these Lubricators you will get value received; and here is a good place to answer the above query. If you have run a threshing engine a season or part of a season you have learned that it is much easier to get a poor grade of oil than a good one, yet your Lubricator will do this at times even with best of oil, and the reason is due to the condition of the feed nozzle at the bottom of the feed glass. The surface around the needle point in the nozzle becomes coated or rough from sediment from the oil. This coating allows the drop to adhere to it until it becomes too large to pass up through the glass without striking the sides and the glass becomes blurred and has the appearance of being full of oil, so in a measure to obviate this Powell's Lubricators are fitted with 3/4 glasses- being of large internal diameter. The permanent remedy however is to take out the glass and clean the nozzle with waste or a rag, rubbing the points smooth and clean. The drop will then release itself at a moderate size and pass up through the glass without any danger of striking the sides. However, if the Lubricator is on crooked it may do this same thing. The remedy is very simple-straighten it up. While talking of the various appliances for oiling your engine you will pardon me if I say that I think every traction engine ought to be supplied with an oil pump as you will find it very convenient for a traction engine especially on the road. For instance, should the engine prime to any great extent your cylinder will require more oil for a few minutes than your sight feed will supply, and here is where, your little pump will help you out. Either the Detroit or Powell people make as good an article of this kind as you can find anywhere, and can furnish you either the glass or metal body.

Hard Grease and a good Cup come next. In my trips over various parts of the country I visit a great many engineers and find a great part of them using hard grease and I also find the quality varying all the way from the very best down to the cheapest grade of axle grease. The Badger Oil I think is the best that can be procured for this purpose, and while I do not know just who makes it, you will probably have but little trouble in finding it, and if you are looking for a first class automatic cup for your wrist pin or crank box get the Wm. Powell Cup from any jobbing supply house.

These people also make a very neat little attachment for their Class "A" Lubricator which is a decided convenience for the engineer, and is called a "Filler." It consists of a second reservoir or cup, of about the same capacity of the reservoir of Lubricator, thus doubling the capacity. It is attached at the filling plug, and is supplied with a fine strainer, which catches all dirt, and grit, allowing only clear oil to enter the lubricator, and by properly manipulating the little shut-off valve the strainer can be removed and cleaned and the cup refilled without disturbing the working of the Lubricator. This little attachment will soon be in general use.

BOILER FEEDERS

Injectors have a dangerous rival in the Moore Steam Pump or boiler feeder for traction engines, and the reason this little pump is not in more general use is the fact that among the oldest methods for feeding a boiler is the independent steam pump and they were always unsatisfactory from the fact that they were a steam engine within themselves, having a crank or disc, flywheel, eccentric, eccentric yoke, valve, valve stem, crosshead, slides, and all the reciprocating parts of a complete engine. Being necessarily very small, these parts of course are very frail and delicate, were easily broken or damaged by the rough usage to which they were subjected while bumping around over rough roads on a traction engine. The Moore Pump, manufactured by The Union Steam Pump Company, of Battle Creek, Mich., is a complete departure from the old steam engine pump, and if you take any interest in any of the novel ways in which steam can be utilized send to them for a circular and sectional cuts and you can spend several hours very profitably in determining just how the direct pressure from the boiler can be made to drive the piston head the full stroke of cylinder, open exhaust port, shift the valve open steam port and drive the piston back again and repeat the operation as long as the boiler pressure is allowed to reach the pump and yet have no connection whatever with any of the reciprocating parts of the pump, and at the same time lift and force water into the boiler in any quantity desired.

Another novel feature in this "little boiler feeder" is that after the steam has acted on the cylinder it can be exhausted directly into the feed water, thus utilizing all its heat to warm the water before entering the boiler. Now it required a certain number of heat units to produce this steam which after doing its work gives back all its heat again to the feed water and it would be a very interesting problem for some of the young engineers, as well as the old ones, to determine just what loss if any is sustained in this manner of supplying a boiler. If you are thinking of trying an independent pump, don't be afraid of this one. I take particular pride in recommending anything that I have tried myself, and know to be as recommended.

And a boiler feeder of this kind has all the advantage of the injector, as it will supply the boiler without running the engine, and it has the advantage over the injector, in not being so delicate, and will work water that can not be handled by the best of injectors.

We have very frequently had this question put to us: "Ought I to grease my gearing?" If I said "yes," I had an argument on my hands at once. If I said "no," some one would disagree just as quickly, and how shall I answer it to the satisfaction of most engineers of a traction engine?

I always say what I have to say and stay by it until I am convinced of the error. Now some of you will smile when I say that the only thing for gear where there is dust, is "Mica Axle Grease." And you smile because you don't know what it is made of, but think it some common grease named for some old saint, but that is not the case. If these people who make this lubricant would give it another name, and get it introduced among engineers, nothing else would be used. You have seen it advertised for years as an axle grease and think that is all it is good for; and there is where you make a mistake. It is made of a combination of solid lubricant and ground or pulverized mica, that is where it gets its name, and nothing can equal mica as a lubricant if you could apply it to your gear; and to do this it has been combined with a heavy grease. This in being applied to the gear retains the small particles of mica, which soon imbed themselves in every little abrasion or rough place in the gearing, and the surface quickly becomes hard and smooth throughout the entire face of the engaging gear, and your gear will run quiet, and if your gearing is not out of line will stop cutting if applied in time.

It will run dry and dust will not collect on the surface of your cogs, and after a coating is once formed it should never be disturbed by scraping the face of the gear, and a very little added from time to time will keep your gear in fine shape. Its name is against it and if the makers would take a tumble to themselves and call it "Mica Oil" or some catchy name and get it introduced among the users of tight gearing, they would sell just as much axle grease and all the grease for gearings.

FORCE FEED OILER

Force feed oiler come next on the list. This is something not generally understood by engineers of traction and farm engines, and accounts for it being so far down the list. But we think it will come into general use within a few years, as an oiler of this kind forces the oil instead of depending on gravity.

The Acorn Brass Works of Chicago make a very unique and successful little oiler which forces a small portion of oil in a spray into the valve and cylinder, and repeats the operation at each stroke of the engine, and is so arranged that it stops automatically as soon as the oil is out of the reservoir; and at once calls the attention of the engineer to the fact, and it can be regulated to throw any quantity of oil desired. Is made for any size or make of engine.

SPEEDER

One of the little things, that every engineer ought to have is a Motion counter or speeder. Of course, you can count the revolutions of your engine, but you frequently want to know the speed of the driven pulley, cylinder for instance: When you know the exact size of engine pulley and your cylinder pulley, and the exact speed of your engine, and there was no such thing as the slipping of drive belt, you could figure the speed of your cylinder, but by knowing this and then applying the speeder, you can determine the loss by comparing the figured speed with the actual speed shown by the speeder. If you have a good speeder you can make good use of it every day you run machinery. If you want one you want the best and there is nothing better than the one made by The Tabor Manufacturing Co., of Philadelphia, Pa. We use no other. You will see their advertisement in the American Thresherman.

SPARK ARRESTER

But one article in the entire list did I find to be sectional, and that was for a spark arrester. These inquiries were all without exception from the wooded country, that is, from a section where it is cheaper to burn wood than coal. There is nothing strange that parties running engines in these sections should ask for a spark arrester, as builders of this class of engines usually supply their engines with a "smoke stack", with little or no reference to safety from fire. This being recognized by some genius in one of our wooded states who has profited by it and has produced a "smoke stack" which is also a "spark arrester." This stack is a success in every sense of the word, and is made for any and all styles of farm and saw mill engines. It is made by the South Bend Spark Arrester Co., of South Bend, Indiana, and if you are running an engine and firing with wood or straw, don't run too much risk for the engineer usually comes in for a big share of the blame if a fire is started from the engine. And as the above company make a specialty of this particular article, you will get something reliable if you are in a section where you need it.

LIFTING JACK

Next comes enquiries for a good lifting Jack.

This would indicate that the boys had been getting their engine in a hole, but there are a great many times when a good Jack comes handy, and it will save its cost many times every season.

Too many engineers forget that when he is fooling around that he is the only one losing time. The facts are the entire crew are doing nothing, besides the outfit is making no money unless running.

You want to equip yourself with any tool that will save time.

The Barth Mfg, Co., of Milwaukee, make a Jack especially adapted to this particular work, and every engine should have a "mascot" in the shape of a lifting Jack.

Now before dropping the subject of "handy things for an engineer," I want to say to the engineer who takes pride in his work, that if you would enjoy a touch of high life in engineering, persuade your boss, if you have one, to get you a Fuller Tender made by the Parson's Band Cutter and Feeder Co., Newton, Iowa, and attach to your engine. It may look a little expensive, but a luxury usually costs something and by having one you will do away with a great deal of the rough and tumble part of an engineers life.

And if you want to keep yourself posted as to what is being done by other threshermen throughout the world, read some good "Threshermen's Home journal." The American Thresherman for instance is the "warmest baby in the bunch." And if anything new under the sun comes out you will find it in the pages of this bright and newsy journal. Keep to the front in your business. Your business is as much a business as any other profession, and while it may not be quite as remunerative as a R. R. attorney, or the president of a life insurance company it is just as honorable, and a good engineer is appreciated by his employer just as much as a good man in any other business. A good engineer can not only always have a job, but he can select his work. That is if there is any choice of engines in a neighborhood the best man gets it.

SOMETHING ABOUT PRESSURE

Now before bringing this somewhat lengthy lecture to a close, (for I consider it a mere lecture, a talk with the boys) I want to say something more about pressure. You notice that I have not advocated a very high pressure; I have not gone beyond 125 lbs. and yet you know and I know that very much higher pressure is being carried wherever the traction engine is used, and I want to say that a very high pressure is no gauge or guarantee of the intelligence of the engineer. The less a reckless individual knows about steam the higher pressure he will carry. A good engineer is never afraid of his engine without a good reason, and then he refuses to run it. He knows something of the enormous pressure in the boiler, while the reckless fellow never thinks of any pressure beyond the 100 or 140 pounds that his gauge shows. He says, "'O! That,' that aint much of a pressure, that boiler is good for 200 pounds." It has never dawned on his mind (if he has one) that that 140 pounds mean 140 pounds on every square inch in that boiler shell, and 140 on each square inch of tube sheets. Not only this but every square inch in the shell is subjected to two times this pressure as the boiler has two sides or in other words, each square inch has a corresponding opposite square inch, and the seam of shell must sustain this pressure, and as a single riveted boiler only affords 62 per cent of the strength of solid iron. It is something that every engineer ought to consider. He ought to be able to thoroughly appreciate this almost inconceivable pressure. How many engineers are today running 18 and 20 horse power engines that realizes that a boiler of this diameter is not capable of sustaining the pressure he had been accustomed to carry in his little 26 or 30 inch boiler? On page 114 You will get some idea of the difference in safe working pressure of boilers, of different diameters. On the other hand this is not intended to make you timid or afraid of your engine, as there is nothing to be afraid of if you realize what you are handling, and try to comprehend the fact that your steam gauge represents less than one 1-1000 part of the power you have under your management. You never had this put to you in this light before, did you?

If you thoroughly appreciate this fact and will try to comprehend this power confined in your boiler by noting the pressure, or power exerted by your cylinder through the small supply pipe, you will soon be an engineer who will only carry a safe and economical pressure, and if there comes a time when it is necessary to carry a higher pressure, you will be an engineer who will set the pop back again, when or as soon as this extra pressure is not necessary.

If I can get you to comprehend this power proposition no student of "Rough and Tumble Engineering" will ever blow up a boiler.

When I started out to talk engine to you I stated plainly that this book would not be filled up with scientific theories, that while they were very nice they would do no good in this work. Now I am aware that I could have made a book four times as large as this and if I had, it would not be as valuable to the beginner as it is now.

From the fact that there is not a problem or a question contained in it that any one who has a common school education can not solve or answer without referring to any textbooks The very best engineer in the country need not know any more than he will find in these pages. Yet I don't advise you to stop here, go to the top if you have the time and opportunity. Should I have taken up each step theoretically and given forms, tables, rules and demonstrations, the young engineer would have become discouraged and would never have read it through. He would have become discouraged because he could not understand it. Now to illustrate what I mean, we will go a little deeper and then still deeper, and you will begin to appreciate the simple way of putting the things which you as a plain engineer are interested in.

For example on page 114 we talked about the safe working pressure of different sized boilers. It was most likely natural for you to say "How do I find the safe working pressure?" Well, to find the safe working pressure of a boiler it is first necessary to find the total pressure necessary to burst the boiler. It requires about twice as much pressure to tear the ends out of a boiler as it does to burst the shell, and as the weakest point is the basis for determining the safe pressure, we will make use of the shell only.

We will take for example a steel boiler 32 inches in diameter and 6 ft. long, 3/8 in. thick, tensile strength 60,000 lbs. The total pressure required to burst this shell would be the area exposed times the pressure. The thickness multiplied by the length then by 2 (as there are two sides) then by the tensile strength equals the bursting pressure: 3/8 x 72 X 2 x 60,000 = 3,240,000 the total bursting pressure and the pressure per square inch required to burst the shell is found by dividing the total bursting pressure 3,240,000 pounds by the diameter times the length 3,240,000 / (32 x 72) = 1406 lbs.

It would require 1406 lbs. per square inch to burst this shell if it were solid, that is if it had no seam, a single seam affords 62 per cent of the strength of shell, 1406 x .62 = 871 lbs. to burst the seam if single riveted; add 20 per cent if double riveted.

To determine the safe working pressure divide the bursting pressure of the weakest place by the factor of safety. The United States Government use a factor of 6 for single riveted and add 20 per cent for double riveted, 871 / 6 = 145 lbs. the safe working pressure of this particular boiler, if single riveted and 145 + 20 per cent=174 double riveted.

Now suppose you take a boiler the same length and of the same material, but 80 inches in diameter. The bursting pressure would be 3,240,000 / (80 x 72) = 560 lbs., and the safe working pressure would be 560 / 6 = 93 lbs.

You will see by this that the diameter has much to do with the safe working pressure, also the diameter and different lengths makes a difference in working pressure.

Now all of this is nice for you to know, and it may start you on a higher course, it will not make you handle your engine any better, but it may convince you that there is something to learn.

Suppose we give you a little touch of rules, and formula in boiler making.

For instance you want to know the percent of strength of single riveted and double riveted as compared to solid iron. Some very simple rules, or formula, are applicable.

Find the percent of strength to the solid iron in a single-riveted seam, 1/4 inch plate, 5/8 inch rivet, pitched or spaced 2 inch centers. First reduce all to decimal form, as it simplifies the calculation; 1/4=.25 and 5/8 inch rivets will require 11/16 inch hole, this hole is supposed to be filled by the rivet, after driving, consequently this diameter is used in the calculation, 11/16 inches=.6875.

First find the percent of strength of the sheet.

The formula is $\dfrac{P-D}{P}$ = percent.

P = the pitch, D = the diameter of the rivet hole, percent = percent of strength of the solid iron.

Substituting values, $\dfrac{2 - .6875}{2}$ = .66.

Now of course you understand all about that, but it is Greek to some people.

So you see I have no apologies to make for following out my plain comprehensive talk, have not confused you, or lead you to believe that it requires a great amount of study to become an engineer. I mean a practical engineer, not a mechanical engineer. I just touch mechanical engineering to show you that that is something else. If you are made of the proper stuff you can get enough out of this little book to make you as good an engineer as ever pulled a throttle on a traction engine. But this is no novel. Go back and read it again, and ever time you read it you will find something you had not noticed before.

The Codes Of Hammurabi And Moses
W. W. Davies

QTY

The discovery of the Hammurabi Code is one of the greatest achievements of archaeology, and is of paramount interest, not only to the student of the Bible, but also to all those interested in ancient history...

Religion　　**ISBN:** *1-59462-338-4*　　　**Pages:132**
MSRP $12.95

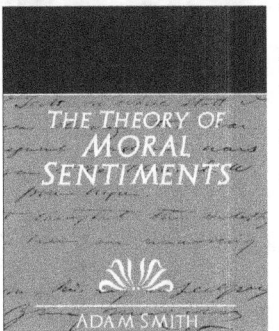

The Theory of Moral Sentiments
Adam Smith

QTY

This work from 1749. contains original theories of conscience amd moral judgment and it is the foundation for systemof morals.

Philosophy　**ISBN:** *1-59462-777-0*　　　**Pages:536**
MSRP $19.95

Jessica's First Prayer
Hesba Stretton

QTY

In a screened and secluded corner of one of the many railway-bridges which span the streets of London there could be seen a few years ago, from five o'clock every morning until half past eight, a tidily set-out coffee-stall, consisting of a trestle and board, upon which stood two large tin cans, with a small fire of charcoal burning under each so as to keep the coffee boiling during the early hours of the morning when the work-people were thronging into the city on their way to their daily toil...

Pages:84

Childrens　　**ISBN:** *1-59462-373-2*　　*MSRP $9.95*

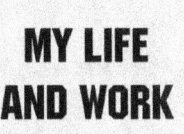

My Life and Work
Henry Ford

QTY

Henry Ford revolutionized the world with his implementation of mass production for the Model T automobile. Gain valuable business insight into his life and work with his own auto-biography... "We have only started on our development of our country we have not as yet, with all our talk of wonderful progress, done more than scratch the surface. The progress has been wonderful enough but..."

Pages:300

Biographies/　　**ISBN:** *1-59462-198-5*　　*MSRP $21.95*

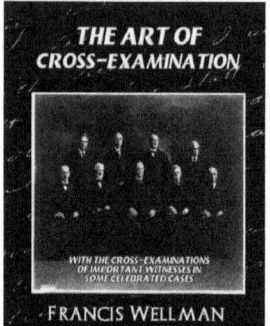

The Art of Cross-Examination
Francis Wellman

QTY

I presume it is the experience of every author, after his first book is published upon an important subject, to be almost overwhelmed with a wealth of ideas and illustrations which could readily have been included in his book, and which to his own mind, at least, seem to make a second edition inevitable. Such certainly was the case with me; and when the first edition had reached its sixth impression in five months, I rejoiced to learn that it seemed to my publishers that the book had met with a sufficiently favorable reception to justify a second and considerably enlarged edition. ..

Reference   ISBN: *1-59462-647-2*

Pages:412

MSRP $19.95

On the Duty of Civil Disobedience
Henry David Thoreau

QTY

Thoreau wrote his famous essay, On the Duty of Civil Disobedience, as a protest against an unjust but popular war and the immoral but popular institution of slave-owning. He did more than write—he declined to pay his taxes, and was hauled off to gaol in consequence. Who can say how much this refusal of his hastened the end of the war and of slavery ?

Law   ISBN: *1-59462-747-9*

Pages:48

MSRP $7.45

Dream Psychology Psychoanalysis for Beginners
Sigmund Freud

QTY

Sigmund Freud, born Sigismund Schlomo Freud (May 6, 1856 - September 23, 1939), was a Jewish-Austrian neurologist and psychiatrist who co-founded the psychoanalytic school of psychology. Freud is best known for his theories of the unconscious mind, especially involving the mechanism of repression; his redefinition of sexual desire as mobile and directed towards a wide variety of objects; and his therapeutic techniques, especially his understanding of transference in the therapeutic relationship and the presumed value of dreams as sources of insight into unconscious desires.

Pages:196

Psychology   ISBN: *1-59462-905-6*

MSRP $15.45

The Miracle of Right Thought
Orison Swett Marden

QTY

Believe with all of your heart that you will do what you were made to do. When the mind has once formed the habit of holding cheerful, happy, prosperous pictures, it will not be easy to form the opposite habit. It does not matter how improbable or how far away this realization may see, or how dark the prospects may be, if we visualize them as best we can, as vividly as possible, hold tenaciously to them and vigorously struggle to attain them, they will gradually become actualized, realized in the life. But a desire, a longing without endeavor, a yearning abandoned or held indifferently will vanish without realization.

Pages:360

Self Help   ISBN: *1-59462-644-8*

MSRP $25.45

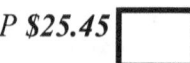

QTY

☐ **The Rosicrucian Cosmo-Conception Mystic Christianity** *by Max Heindel*   ISBN: *1-59462-188-8*   **$38.95**
The Rosicrucian Cosmo-conception is not dogmatic, neither does it appeal to any other authority than the reason of the student. It is: not controversial, but is: sent forth in the, hope that it may help to clear...   New Age/Religion Pages 646

☐ **Abandonment To Divine Providence** *by Jean-Pierre de Caussade*   ISBN: *1-59462-228-0*   **$25.95**
"The Rev. Jean Pierre de Caussade was one of the most remarkable spiritual writers of the Society of Jesus in France in the 18th Century. His death took place at Toulouse in 1751. His works have gone through many editions and have been republished...   Inspirational/Religion Pages 400

☐ **Mental Chemistry** *by Charles Haanel*   ISBN: *1-59462-192-6*   **$23.95**
Mental Chemistry allows the change of material conditions by combining and appropriately utilizing the power of the mind. Much like applied chemistry creates something new and unique out of careful combinations of chemicals the mastery of mental chemistry...   New Age Pages 354

☐ **The Letters of Robert Browning and Elizabeth Barret Barrett 1845-1846 vol II**   ISBN: *1-59462-193-4*   **$35.95**
by Robert Browning and Elizabeth Barrett   Biographies Pages 596

☐ **Gleanings In Genesis (volume I)** *by Arthur W. Pink*   ISBN: *1-59462-130-6*   **$27.45**
Appropriately has Genesis been termed "the seed plot of the Bible" for in it we have, in germ form, almost all of the great doctrines which are afterwards fully developed in the books of Scripture which follow...   Religion/Inspirational Pages 420

☐ **The Master Key** *by L. W. de Laurence*   ISBN: *1-59462-001-6*   **$30.95**
In no branch of human knowledge has there been a more lively increase of the spirit of research during the past few years than in the study of Psychology, Concentration and Mental Discipline. The requests for authentic lessons in Thought Control, Mental Discipline and...   New Age/Business Pages 422

☐ **The Lesser Key Of Solomon Goetia** *by L. W. de Laurence*   ISBN: *1-59462-092-X*   **$9.95**
This translation of the first book of the "Lernegton" which is now for the first time made accessible to students of Talismanic Magic was done, after careful collation and edition, from numerous Ancient Manuscripts in Hebrew, Latin, and French...   New Age/Occult Pages 92

☐ **Rubaiyat Of Omar Khayyam** *by Edward Fitzgerald*   ISBN: *1-59462-332-5*   **$13.95**
Edward Fitzgerald, whom the world has already learned, in spite of his own efforts to remain within the shadow of anonymity, to look upon as one of the rarest poets of the century, was born at Bredfield, in Suffolk, on the 31st of March, 1809. He was the third son of John Purcell...   Music Pages 172

☐ **Ancient Law** *by Henry Maine*   ISBN: *1-59462-128-4*   **$29.95**
The chief object of the following pages is to indicate some of the earliest ideas of mankind, as they are reflected in Ancient Law, and to point out the relation of those ideas to modern thought.   Religion/History Pages 452

☐ **Far-Away Stories** *by William J. Locke*   ISBN: *1-59462-129-2*   **$19.45**
"Good wine needs no bush, but a collection of mixed vintages does. And this book is just such a collection. Some of the stories I do not want to remain buried for ever in the museum files of dead magazine-numbers an author's not unpardonable vanity..."   Fiction Pages 272

☐ **Life of David Crockett** *by David Crockett*   ISBN: *1-59462-250-7*   **$27.45**
"Colonel David Crockett was one of the most remarkable men of the times in which he lived. Born in humble life, but gifted with a strong will, an indomitable courage, and unremitting perseverance...   Biographies/New Age Pages 424

☐ **Lip-Reading** *by Edward Nitchie*   ISBN: *1-59462-206-X*   **$25.95**
Edward B. Nitchie, founder of the New York School for the Hard of Hearing, now the Nitchie School of Lip-Reading, Inc, wrote "LIP-READING Principles and Practice". The development and perfecting of this meritorious work on lip-reading was an undertaking...   How-to Pages 400

☐ **A Handbook of Suggestive Therapeutics, Applied Hypnotism, Psychic Science**   ISBN: *1-59462-214-0*   **$24.95**
by Henry Munro   Health/New Age/Health/Self-help Pages 376

☐ **A Doll's House: and Two Other Plays** *by Henrik Ibsen*   ISBN: *1-59462-112-8*   **$19.95**
Henrik Ibsen created this classic when in revolutionary 1848 Rome. Introducing some striking concepts in playwriting for the realist genre, this play has been studied the world over.   Fiction/Classics/Plays 308

☐ **The Light of Asia** *by sir Edwin Arnold*   ISBN: *1-59462-204-3*   **$13.95**
In this poetic masterpiece, Edwin Arnold describes the life and teachings of Buddha. The man who was to become known as Buddha to the world was born as Prince Gautama of India but he rejected the worldly riches and abandoned the reigns of power when...   Religion/History/Biographies Pages 170

☐ **The Complete Works of Guy de Maupassant** *by Guy de Maupassant*   ISBN: *1-59462-157-8*   **$16.95**
"For days and days, nights and nights, I had dreamed of that first kiss which was to consecrate our engagement, and I knew not on what spot I should put my lips..."   Fiction/Classics Pages 240

☐ **The Art of Cross-Examination** *by Francis L. Wellman*   ISBN: *1-59462-309-0*   **$26.95**
Written by a renowned trial lawyer, Wellman imparts his experience and uses case studies to explain how to use psychology to extract desired information through questioning.   How-to/Science/Reference Pages 408

☐ **Answered or Unanswered?** *by Louisa Vaughan*   ISBN: *1-59462-248-5*   **$10.95**
Miracles of Faith in China   Religion Pages 112

☐ **The Edinburgh Lectures on Mental Science (1909)** *by Thomas*   ISBN: *1-59462-008-3*   **$11.95**
This book contains the substance of a course of lectures recently given by the writer in the Queen Street Hail, Edinburgh. Its purpose is to indicate the Natural Principles governing the relation between Mental Action and Material Conditions...   New Age/Psychology Pages 148

☐ **Ayesha** *by H. Rider Haggard*   ISBN: *1-59462-301-5*   **$24.95**
Verily and indeed it is the unexpected that happens! Probably if there was one person upon the earth from whom the Editor of this, and of a certain previous history, did not expect to hear again...   Classics Pages 380

☐ **Ayala's Angel** *by Anthony Trollope*   ISBN: *1-59462-352-X*   **$29.95**
The two girls were both pretty, but Lucy who was twenty-one who supposed to be simple and comparatively unattractive, whereas Ayala was credited, as her Bombwhat romantic name might show, with poetic charm and a taste for romance. Ayala when her father died was nineteen...   Fiction Pages 484

☐ **The American Commonwealth** *by James Bryce*   ISBN: *1-59462-286-8*   **$34.45**
An interpretation of American democratic political theory. It examines political mechanics and society from the perspective of Scotsman James Bryce   Politics Pages 572

☐ **Stories of the Pilgrims** *by Margaret P. Pumphrey*   ISBN: *1-59462-116-0*   **$17.95**
This book explores pilgrims religious oppression in England as well as their escape to Holland and eventual crossing to America on the Mayflower, and their early days in New England...   History Pages 268

QTY

The Fasting Cure *by Sinclair Upton*   ISBN: *1-59462-222-1*   **$13.95**
In the Cosmopolitan Magazine for May, 1910, and in the Contemporary Review (London) for April, 1910, I published an article dealing with my experiences in fasting. I have written a great many magazine articles, but never one which attracted so much attention... New Age/Self Help/Health Pages 164

Hebrew Astrology *by Sepharial*   ISBN: *1-59462-308-2*   **$13.45**
In these days of advanced thinking it is a matter of common observation that we have left many of the old landmarks behind and that we are now pressing forward to greater heights and to a wider horizon than that which represented the mind-content of our progenitors...   Astrology Pages 144

Thought Vibration or The Law of Attraction in the Thought World   ISBN: *1-59462-127-6*   **$12.95**

by William Walker Atkinson   *Psychology/Religion Pages 144*

Optimism *by Helen Keller*   ISBN: *1-59462-108-X*   **$15.95**
Helen Keller was blind, deaf, and mute since 19 months old, yet famously learned how to overcome these handicaps, communicate with the world, and spread her lectures promoting optimism. An inspiring read for everyone...   Biographies/Inspirational Pages 84

Sara Crewe *by Frances Burnett*   ISBN: *1-59462-360-0*   **$9.45**
In the first place, Miss Minchin lived in London. Her home was a large, dull, tall one, in a large, dull square, where all the houses were alike, and all the sparrows were alike, and where all the door-knockers made the same heavy sound...   Childrens/Classic Pages 88

The Autobiography of Benjamin Franklin *by Benjamin Franklin*   ISBN: *1-59462-135-7*   **$24.95**
The Autobiography of Benjamin Franklin has probably been more extensively read than any other American historical work, and no other book of its kind has had such ups and downs of fortune. Franklin lived for many years in England, where he was agent...   Biographies/History Pages 332

Name	
Email	
Telephone	
Address	
City, State ZIP	

☐ **Credit Card**   ☐ **Check / Money Order**

Credit Card Number	
Expiration Date	
Signature	

Please Mail to: Book Jungle
PO Box 2226
Champaign, IL 61825
or Fax to: 630-214-0564

ORDERING INFORMATION
web: *www.bookjungle.com*
email: *sales@bookjungle.com*
fax: *630-214-0564*
mail: *Book Jungle PO Box 2226 Champaign, IL 61825*
or PayPal *to sales@bookjungle.com*

Please contact us for bulk discounts

DIRECT-ORDER TERMS

**20% Discount if You Order
Two or More Books**
Free Domestic Shipping!
Accepted: Master Card, Visa,
Discover, American Express